RADIO!

G. HOWARD POTEET

Pflaum Publishing
Dayton, Ohio

Acknowledgment is gratefully made of permission to use the following materials:

Radio script "Channel Cat in the Middle Distance" by Jean Shepherd. Used with permission.

Gangbusters, "The Eddie Doll Case" (1939, sponsored by Colgate-Palmolive-Peet Co., written by Brice Disque, Jr.). Used with permission of Charles Michelson Incorporated.

Quiz Kids script. Used with permission of Louis G. Cowan.

Jack Armstrong script and Wheaties commercial (p. 54) used with the permission of General Mills, Inc.

Ma Perkins script and Crisco (p. 96) and Lava (p. 114) commercials used with permission of Procter & Gamble Company.

Saturday Night Swing Club used with permission of CBS Radio.

Newscast transcript by Lowell Thomas. From *History As You Heard It* by Lowell Thomas. ©1957. Used with permission of Doubleday & Co., Inc.

Pepsi-Cola commercial (p. 3). Reproduced by permission of Pepsi-Cola Company ©1940 Pepsi-Cola Company.

Lifebuoy commercial (p. 7) and Rinso commercial (p. 68) used with permission of Lever Brothers Company.

Halo commercial (p. 35) and Super Suds commercial (p. 79) used with permission of Colgate-Palmolive Company.

Barbasol commercial (p. 104) used with permission of Leeming/Pacquin, division of Pfizer, Inc.

Gillette commercial (p. 109). Reprinted by permission of The Gillette Company. All rights reserved.

The Aldrich Family used with permission of Mrs. Clifford Goldsmith and the William Morris Agency, Inc.

ISBN 0-8278-0227-7
Library of Congress Catalog Card Number 75-14640
Book Design: Joe Loverti
Cover: adapted from a photo by H. Armstrong Roberts.

Contents

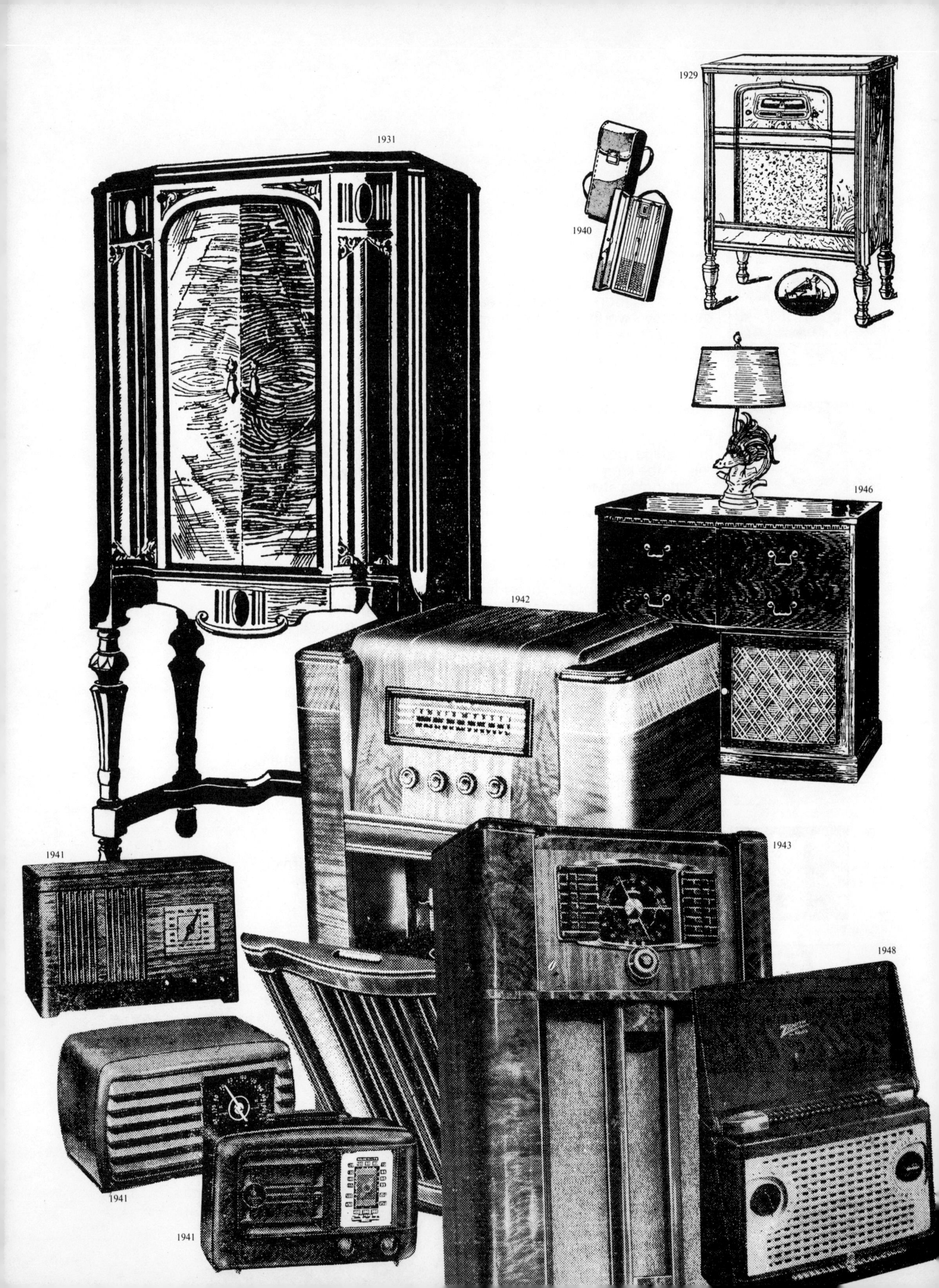
1929
1931
1940
1946
1942
1941
1943
1948
1941
1941

Introduction

What does radio mean to you? Cheap pocket-sized transistor sets? Music and news programs all day long? Phone-in shows where anyone can say what's on his mind?

It wasn't always that way. At one time people paid as much as $1,000 for a radio. Drama took up more radio time than did music and news. And only tuxedo-clad professional announcers who never mispronounced a word spoke before the mike.

Those early radio sets were enormous monsters with wires and tubes everywhere. Kids made smaller, cheaper models from boards, round boxes, and a device called a "cat's whisker." With both expensive sets and homemade ones, you listened to earphones clamped over your ears.

And what did you hear? Mostly static, squealing, whistling, and howling noises. And sometimes music. Since radio signals were stronger at night, you might stay up to midnight and be lucky enough to hear a faint voice giving the call letters of a station in the next state. That was in the early 1920s.

During the last part of the 1920s and in the 1930s, radio took America by storm. People loved the free entertainment that came into their homes. "Wanna buy a duck?" and other popular phrases were repeated by people who heard them on radio. You remember that *Laugh-in* gave us similar expressions.

Radio programs entranced listeners. Each night at many movie houses, the film was interrupted so that the audience could listen to a popular show, *Amos 'n' Andy.* When a radio star endorsed a product, people rushed to buy it.

At night, kids dashed home in order not to miss an episode of their favorite radio serial like *Superman* or *Jack Armstrong.* Mothers wept over the troubles of soap opera heroines like Helen Trent or Stella Dallas. Fathers pondered the weighty words of newsmen like Lowell Thomas or H. V. Kaltenborn. Radio took up more and more of the average person's time.

After 1946, when television became available, fewer people listened to the radio. As a result, only record shows and news were broadcast. This lasted through the late forties, the fifties, and into the early sixties.

In the past ten years, changes in radio programming have been many. The radio listener now can hear all-day news programs, rock music concerts, and participate in phone-in talk shows.

Newer things are yet to come. It's hard to guess what because it is impossible to predict the future. However, it is fun to look at the past—especially if the past is as interesting and wacky as old time radio. That's what this book does. The following chapters look at some of the programs that were popular and explains why your parents couldn't wait to get home to their radio.

1931

Chapter 1

Dit Dit . . . dah dah dah . . .
The first signals transmitted by radio or radio-telegraphy, as it was then called, were in code, that is, in a series of dots and dashes. At first, all communications traveled through wire—by telegraph or telephone. Radio made communication possible without wires; thus, it was frequently called "wireless."

Marconi and the Wireless and Others
Although an eccentric inventor, Nathan B. Stubblefield, transmitted his voice through the air in 1892, he never achieved fame or credit (probably he shunned publicity because he feared someone would steal his idea). Most authorities credit Guglielmo Marconi with transmitting the first wireless signal in 1895. But it is not unusual for two inventors working independently to arrive at the same idea.

Four years later, employed by *The New York Herald*, Marconi reported the America Cup Race and, further, demonstrated his invention for the armed forces. However, his transmissions were in code—not voice. On December 12, 1901, Marconi transmitted the letter "S" in code from Poldhu, Cornwall across the Atlantic to a radio receiver manned by Percy Paget in St. Johns, Newfoundland. People raved about this incredible feat.

But others developed ideas too. Reginald Fessenden (a University of Pittsburgh professor), Lee DeForest and John Ambrose Fleming were three inventors who developed the ideas which made radio practical. Fessenden, asked by the Weather Bureau to test transmitting weather reports, developed a system of continuous wave broadcasts (on which a voice was modulated or superimposed as variations) in contrast to Marconi's transmission of bursts. On Christmas Eve 1906, he succeeded in transmitting a reading of a poem and violin music from Brant Rock, Massachusetts.

Fleming invented a glass detector tube while DeForest (whom some authorities credit with inventing the term *broadcast*) invented the *Audion*, a three element electron tube. He, too, broadcast voice transmissions from the Eiffel Tower in 1908, heard some 500 miles away. Lee DeForest competed with Marconi and another experimenter to broadcast the results of a boat race in 1901 but failed. All three transmitted on the same frequency, creating gibberish.

Secret Radio?
Their energies were not aimed towards developing a medium for entertainment but rather a device that could transmit information: messages, weather reports, and army movements. Surprisingly, many people criticized early radio as a communication device because it was *not* private—conversations could be overheard. There were inventions which attempted to overcome this. However, Lee DeForest seems to have been the first to realize that this apparent shortcoming could be exploited and that the wireless could aid communication by distributing information to a large audience simultaneously.

First Broadcasts
KQW (San Jose, California) claims to have made its first broadcast in 1909 and ran regularly scheduled programs in 1912.

In 1910 DeForest arranged for Enrico Caruso, a famous operatic tenor, to broadcast from the Metropolitan Opera House in New York City. Radio listeners were shocked and startled to hear voices and music instead of time signals and Morse Code.

RADIO FACTS!

How Radio Works!

1. Sound waves striking the microphone are changed into electrical waves which travel to the transmitter.
2. The transmitter changes these electrical waves into radio waves which are transmitted from the antenna.
3. The antenna of a receiver picks up the waves.
4. A tuner in the receiver selects the desired frequency or station. Radio waves are changed back into electrical waves.
5. An amplifier makes the signal stronger and a loudspeaker changes the electrical waves back into sound waves.

and now a word from our sponsor . . .

Pepsi-Cola hits the spot.
Twelve full ounces—that's a lot.
Twice as much for a nickel, too.
Pepsi-Cola is the drink for you.

1929

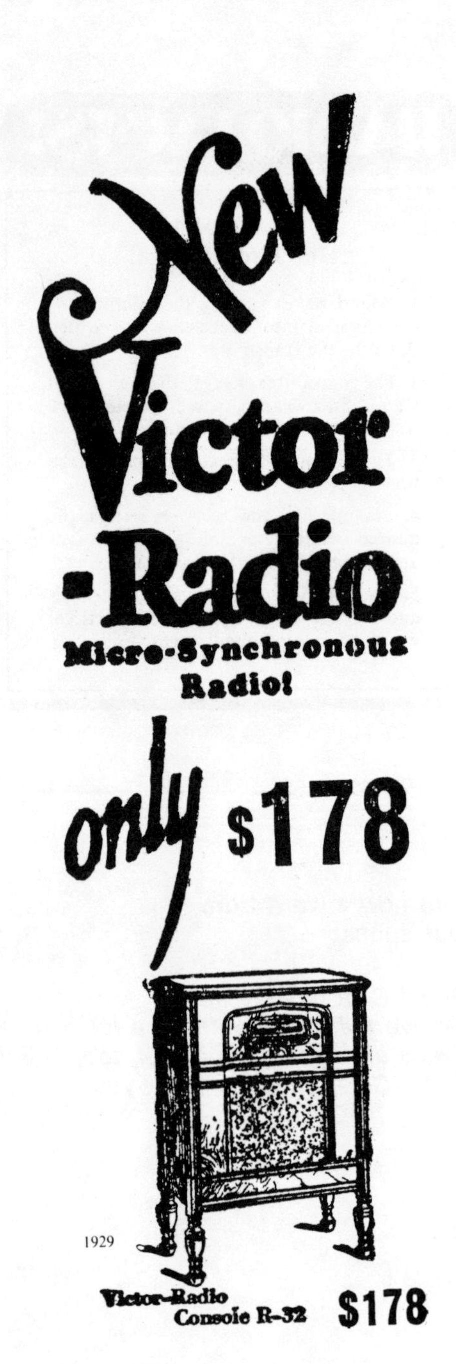

By 1912 many young people were interested in building radios, although they had difficulty getting parts (there were only two distributors of radio parts in the entire country). Nonetheless, they began to build transmitters and receivers and communicated with each other. Sparks flew, arcs popped, and electricity crackled as experimenters worked and struggled with their equipment. Politicians looking towards the future began to see that some control was needed; the air was becoming cluttered with broadcasts. The Navy, a pioneer in radio, was beseiged with practical jokers, crank radio messages, and electrical interference. The result was the Radio Act of 1912. At the time this law was passed, there were 600 amateur stations, 123 experimental stations on land, and 405 ships equipped with wireless. This act required that these transmitters be licensed. (George H. Lewis, Cincinnati, Ohio got the first license.) The statute also divided the radio bands into assigned wave lengths.

David Sarnoff and The Titanic

In 1912 David Sarnoff (later to become President of RCA) manning an early wireless station heard distress signals sent out by the sinking SS Titanic. He relayed the tragic message to the rest of the world. Events like this made the world realize that radio was a vital communication tool.

By 1914 the number of licensed stations had increased to 2,000. Two years later the first nation-wide relay (in which a message was received and then repeated) took place on February 22, 1916 with a message sent to every governor in the United States. Election returns were also broadcast for the first time (by DeForest) in 1916 giving the wrong results: Charles Evans Hughes was declared President. Radio was growing but still was not available to most people.

David Sarnoff, now an American Marconi Company employee, suggested to his bosses than an inexpensive "radio music box" could be made with different channels for the reception of music or news. This cheap device, he believed, would sell well. However, it wasn't produced until 1921. (In 1906 a $7.50 radio had been sold by Telimco which had a range of one mile.)

Patent Troubles

Westinghouse, General Electric, and others began to patent numerous radio devices to protect their interests. So many lawsuits were filed that it looked as if radio companies would stop production.

War began in 1917 (the United States against Germany) and by government order radio hams were forced to stop broadcasting. 3,500 of them joined the service of the United States. The war gave impetus to improving radio; under pressure to improve communications, the government promised legal protection for new inventions. It was a year after the war ended that broadcasting began again. Legal problems again became threatening and were finally solved by the creation of a large company, RCA, which bought the rights to market inventions and made agreements with every company in the radio business.

Early Broadcasts

Dr. Frank Conrad in 1920 operated Station 8XK for Westinghouse in Pittsburgh. He began playing records and reading over the air. Hornes, a department store in Pittsburgh, published an ad suggesting that people buy sets so that they could listen to him.

Other stations went on the air with continuous programming. 8MK (WWJ later) began broadcasting on September 1920 in Detroit, Michigan. On November 2, 1920, the 100 watt station KDKA started broadcasting from the roof of a Pittsburgh Westinghouse plant. The first program was the Harding Election returns. In Newark, New Jersey, WJZ went on the air in 1921, broadcasting the Giants and Yankees in the World Series. By 1921 many stations—WOR (Newark), KYW (Chicago), WWJ (Detroit)—began to broadcast regularly and announced their schedules so that radio owners would know when to tune them in.

1930

1930

The "Music Box"

In 1921 power tubes became readily available, making possible the music box that Sarnoff (now the head of RCA) had dreamed of. Costing $75 each, sales of the set, named the Radiola, reached eleven million dollars in 1922 (the first year of production). By 1923 sales were twenty-two million and in 1924 fifty million dollars. In 1925, the year the International Amateur Radio Union was formed (April 2, 1925), radio had swept the country.

Most built their own sets, however, and they usually were crystal sets using a piece of germanium (a rock highly sensitive to radio waves) and a coiled copper wire which faintly resembled a cat-whisker (which was what it was promptly dubbed). The cat-whisker was twisted until a sensitive spot was found on the crystal (hooked to a 100' long outside antenna wire and grounded by a connection to a cold water pipe).

In order to tune in and separate stations, an oatmeal box was wrapped with wire, attached to the crystal, and a strap arranged to rub against it at various spots to tap the connection at different lengths of wire or wavelengths.

The whole device was fixed to a plain wooden board (called a breadboard because many of them actually were). An earphone was connected and the builder started listening. The materials were cheap; the most expensive part was the headphone at $1.98. Often families would place the ear-pinching headphone in a deep dish and sit around it with their heads together and their ears almost in the bowl so that they could all hear at once. Thus, a soup bowl became the first loudspeaker.

And what did radio nuts hear? They listened to Morse Code signals, to voices reading news from newspapers, to records playing on a phonograph pushed up close to a distant microphone, to piano and violin music. Static distorted the sounds; hums, squeaks, and crackles were louder than the programs which seemed to fade away and then roar back!

But the program content was unimportant. What everyone really wanted to hear was the station's Call Letters so he could gossip about the distance his set had reached the night before.

Radio Entertainment

On August 3, 1922, the first radio drama, *The Wolf*, by Eugene Walter, was broadcast over WGY. Most radio programming was music and soon musicians' unions began to demand royalties for broadcasts of recordings.

In 1923, the year that Zworykin demonstrated a TV system, variety programs began. *The Eveready Hour*, one of the first, was broadcast on WEAF on December 4, 1923.

An early broadcasting station.

The following is what a 1920 broadcast might have sounded like through earphones clamped tightly over your ears.

AWK . . . RAACK . . . hello out there . . . hello ouBROOIEE . . . CRACKLE-SNAP . . . weee . . . this IS . . . ROAR-HISS . . . STAtion 8xkCRASH-hiss ROArrrr . . . pop!

The size of the radio audience increased as more sets were sold. After all, radio was free, wasn't it? All you had to do was turn the dial and you were entertained for free, right?

Well, in a way, yes. Manufacturers paid to advertise their products. In a few years, WEAF was able to charge them $750 an hour for broadcast time. However, until 1924, only WEAF carried commercial advertising; it claimed that right because it was owned by the telephone company (owner of the lines which led from radio stations to transmitters).

WEAF (called a "toll station" because it charged for radio time) on August 28, 1922, charged $50 for a ten-minute commercial designed to sell condominium apartments in Jackson Heights, New York. It was a successful sales pitch.

Most people didn't want radio financed by commercials. Even as late as 1929, the National Association of Broadcasters agreed that between 7:00 and 11:00 p.m., no commercials could be broadcast.

Radio Oddities

In the early twenties, people searching for a scapegoat began to blame their misfortunes on radio. Newspapers carried stories about farmers who complained that radio waves stopped their cows from giving milk or were souring the milk. Some humans claimed to be receiving broadcasts through the fillings in their teeth. Since the body does pick up radio signals (at 60MHz), some of them may have been right.

"Silent Night" was another oddity unkown to most people today but in the period between 1922 and 1927, everybody eagerly awaited it. Local stations did not broadcast that evening so that listeners could tune in stations from out of town. This was necessary because originally all stations were licensed to operate on the same spot on the dial (3600 meters or 833.3 Mhz).

Network Broadcasts

Network broadcasting dates from January 4, 1923, when WEAF and WNAC (Boston) carried the same program at the same time.

The first really important program to be broadcast on a network was in 1924. That year Calvin Coolidge made the first presidential campaign speech on a nationwide hook-up on twenty-six stations (then called a "chain") put together by A.T.&T. In 1925, the inauguration was carried on the same stations.

In 1925 announcers were first permitted to use their names instead of initials. For example, on WEAF, Niles T. Granlund was identified as N. T. G. Apparently, radio fans wanted to know who the announcers were.

The Networks

1926 saw the creation of the National Broadcasting Company. A.T. & T. sold WEAF to RCA for one million dollars. Claiming that radio was "telephony" and thus part of their property, the telephone system had previously refused to permit RCA to use their phone lines or to sell radio time. Now, they were out of the radio/telephony business. Thus, in January 1927, NBC was formed, composed of two networks—the Red (WEAF) and the Blue (WJZ).

On January 27, 1927, a competitor, United Independent Broadcasters, was formed with sixteen stations. In April 1927 a business arrangement was made with Columbia Records in order to get talent. Later the radio company's name was changed to Columbia and then to CBS. In 1928 William Paley took over as its president. (The first TV drama *The Queen's Messenger* was presented that year on W2XAD.)

Four stations (WOR, New York; WXYZ, Detroit; WLW, Cincinnati; and WGN in Chicago) formed the Mutual Broadcasting System in 1934 eventually getting 500 stations into its hookup.

There were other changes. Court decisions were the reason the old Blue Network of NBC became the American Broadcasting Company in 1943. But the history of radio is better seen in its programming as is discussed in the following chapters.

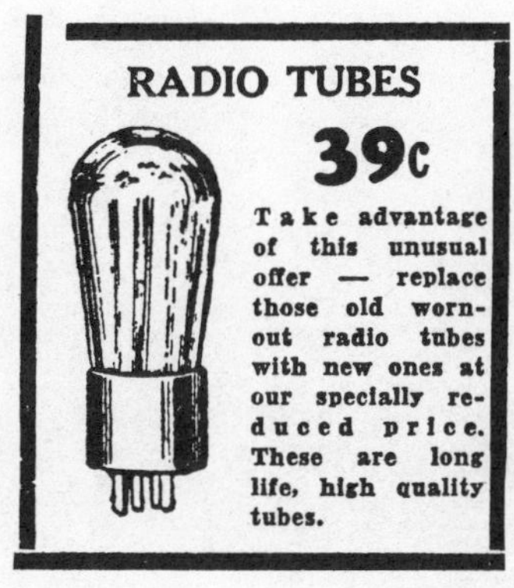

1931

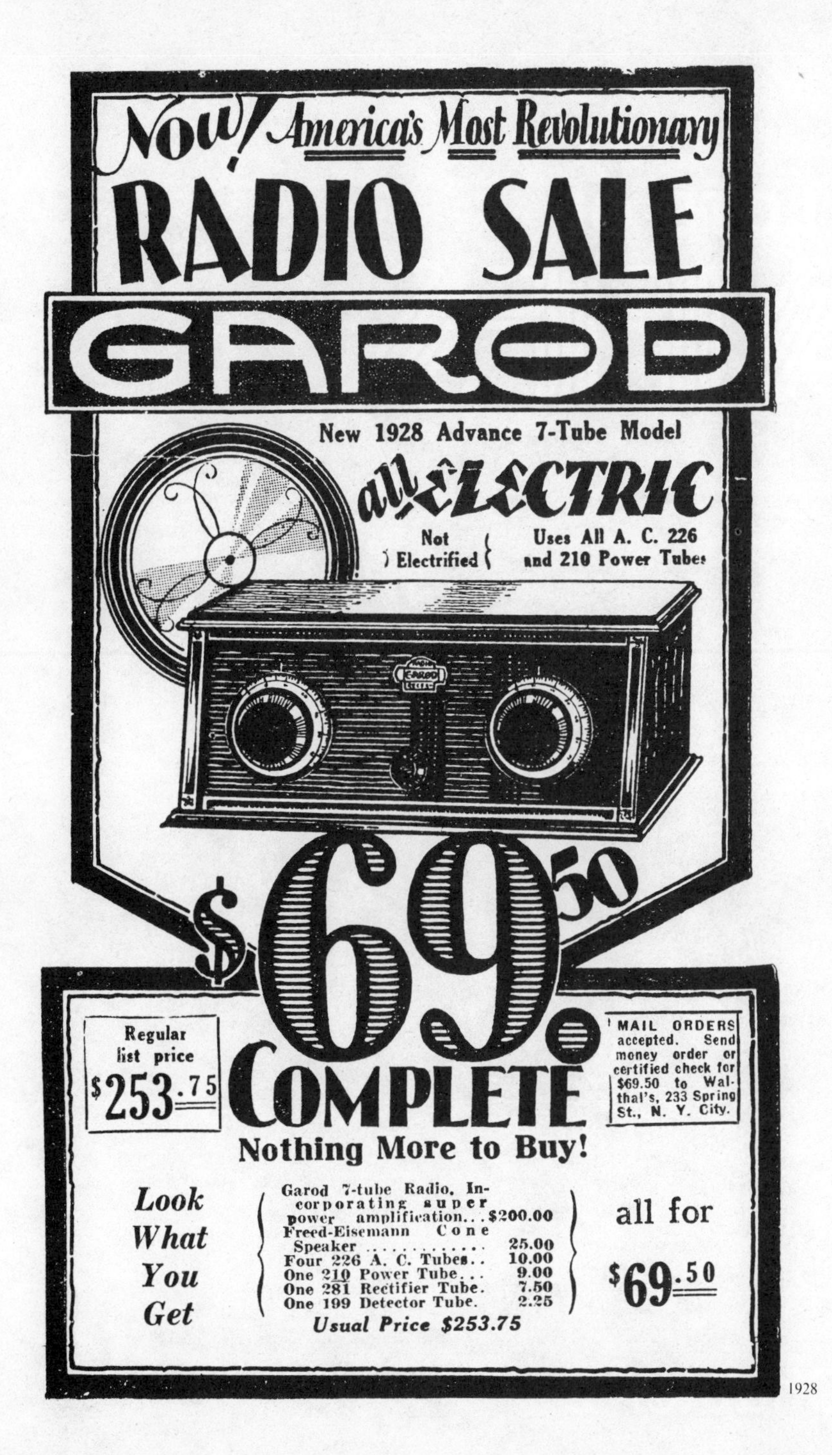

1928

Lifebuoy really stops
(Foghorn)
Beee—oooh!

LOTS OF LAFFS

Chapter 2

Haw! Haw! Oops. Radio was no joke to comedians—vaudeville stars were afraid of it. In the 1920's comics traveled from one booking to another doing the same jokes for two shows a day. A brief spot on just one radio show exposed a gagster's lifetime accumulation of funny stuff to a wider audience than he ever dreamed of. "I can't go on radio—my audience will be lost," one moaned. "Everyone will have heard my act—I won't get any more bookings." Further, jokesters were afraid of other comedians stealing their routines off the air. You see, radio offered them only problems.

Worse, vaudeville teams who did make the jump had the worry of coming up with new material for each new broadcast. George Burns and Gracie Allen were one pair who solved that problem by creating the situation comedy, a formula that still lives on.

Early Radio Comics

The first comedians who succeeded on radio were Colonel Lemuel Q. Stoopnagle and Budd, the Gold Dust Twins, The Tasty Yeast Jesters (a singing group) and, of course, Amos 'n' Andy.

Many stage acts simply couldn't transfer to radio. Sight gags wouldn't do for radio, for radio humor was based on words: mispronunciation, misuse, dialects, puns. Further, timing was more important on radio than on the vaudeville stage. It wasn't easy to be funny on radio.

Goodman Ace was (and is) a master of what is verbally funny; he wrote Jane Ace's malapropisms for radio's *Easy Aces.* From 1930 through 1948 listeners heard Jane say things like, "She got up at the crank of dawn," or "We're living in squander."

Writers Needed

Since radio ate up gags and gagsters at such a fast rate, gag-writers were in demand. Although a stand-up comedian simply told gags that sounded as if they suddenly popped into his head, like Henny Youngman's famous line, "Take my wife . . . please!", one-liners strung together were boring, so skits like those in vaudeville and burlesque were used. Good joke writers found easy employment.

Clever Characterization

Sometimes comedy writers developed one central character. As the late Jack Benny said in a TV interview, the comic character often developed without planning. When something worked, it was repeated. For example, Benny's supposed stinginess got laughs when it was chanced upon; future writers wrote more and more jokes about it into their scripts. (One of Benny's biggest laughs came when a hold-up man said, "Your money or your life." There was a long pause and then the hold-up man said, "Well?" Benny replied, "I'm thinking it over.")

Benny began with a guest spot on the May 2, 1931 *Ed Sullivan Show* and then got his own program in 1932. His show subsisted on a series of running gags such as his Maxwell (an old car), his money vault, and his purported lack of skill in playing the violin.

Although he was selected as the best radio comedian of 1934, Benny was fired by his sponsor. Two years later he was fired again because his sponsor didn't think that he was funny. Benny's humor did not depend upon puns but on situations. He differed from the other radio comics, as Fred Allen pointed out; radio comedy owed a lot to Jack Benny's innovations, such as being the butt of jokes by his stooges, and giving the impression of intimacy—you thought that

1

2

3

4

1 Gracie Allen in a 1937 broadcast. — (NBC photo)

2 George Burns, who had a weekly show *(Burns and Allen)* on NBC — Red. NBC had two networks, Red and Blue. Court action resulted in NBC — Blue becoming ABC. — (NBC photo)

3 Jack Benny and Mary Livingston

4 Rochester starred on the Benny show.

5

you were tuning in Benny at his own home, instead of hearing a man standing on a stage (which was the impression most comics gave.) In addition, it was Benny who introduced satire to radio with his versions of famous movies or stage plays like *Wuthering Heights.*

Sometimes characterizations needed to be thought out and changed. The comedian, George Burns noted that after he and Gracie Allen had been together on stage for many years, they found they were not doing well. Neither they nor their friends could figure out the trouble. Burns finally decided that Gracie was too old for the types of jokes that she was doing (she was playing a young innocent girl). They changed the act so that future gags fitted her age. For example, in one script she shortened all the electric cords in the house to save on electricity.

Catch-Phrases

There were many comedians in the early days who introduced phrases which everybody repeated. One was Joe Penner whose catch-phrase "Wanna buy a duck?" swept the nation. Jack Pearl (Baron Munchausen) affected a German accent which seemed funny to many people because he exaggerated its gutteral sounds which are not used in English. Pearl's famous "Vas you dere Sharlie?" was used whenever anyone doubted his veracity. "Sooo" was the single word catch-phrase of Ed Wynn, the Texaco Fire Chief who was the first comedian to insist on a live audience so he could get a response to his jokes.

Women Comics

The greatest radio comedienne was Fanny Brice who created a character called Baby Snooks, a brat who caused her parents and kid brother Robespierre all kinds of trouble. At one point she put him, instead of his bottles, in the sterilizer. Baby Snooks first appeared on *Follies of the Air*, and she was such a hit she got a weekly program of her own. There were few other women comics: Gracie Fields, Martha Raye, Beatrice Lillie.

Eddie Cantor

To be a radio jester in the late twenties and early thirties was to reach the top in show business, but no comedienne was as successful as any single comedian. The top male comic in those years was a pop-eyed singer and dancer from vaudeville, Eddie Cantor, who really hit it big in radio with his own program after he debuted on Rudy Valee's show in 1931. "Banjo-Eyes" had programs sponsored by a slew of products: Chase and Sanborn, Pebecco, and then Texaco. His song hits included "If You Knew Susie" and "Makin' Whoopie."

His contemporary, Fred Allen, frequently critized him, claiming Cantor would do anything to get a laugh from his studio audience while his listeners were often mystified as to what was going on. Cantor denied this, claiming that he planned his show for the "at-home" audience and wore funny clothes to "make it up" to his studio audience.

It is difficult to see how this is so. On one show, for example, he came on dressed as a bridge lamp. Why? A famous scientist had found the human body generates electricity. Said Cantor, "I can light up the room with one finger." The announcer puzzled, "How?" "Easy," replied Cantor, "I press the switch on the wall." Because electricity was still a novelty, such jokes brought howls of laughter.

5 Jack Pearl (NBC photo)

However, it seems that the studio audience laughed because Cantor's costume was also extremely funny—and the home audience couldn't see it.

Fred Allen

Fred Allen (a juggler turned comedian) was the coiner of many unforgettable phrases ("Hollywood is a great place if you're an orange."). Because he had a gift for turning a clever phrase, he was concerned with how his listening audience would respond on the basis of what they heard, not what the studio audience saw. It is true, however, that one of his funniest shows was visual and that burned him up. An eagle escaped from its trainer and soared around the studio while people screamed and laughed. After the show, Allen was furious because the audience at home had little or no idea of what had happened.

Although his earliest show was called *The Linit—Bath Club* (which began on October 23, 1932), Allen created *Town Hall* and *Allen's Alley* (which began on December 13, 1942) with a regular cast that listeners eagerly awaited each week. For example, there were Senator Claghorn (Kenny Delmar), Mrs. Nussbaum (Minerva Pious), Titus Moody (Parker Fennelly), and Ajax Cassidy (Peter MacDonald), all of whom had distinctive voices and personalities and who always gave uproariously funny answers to the questions that Allen posed to them each week. For example, when Titus Moody was asked if he had a radio, he replied, "No, I don't hold with furniture that talks."

There were lots of zany stooges. Eddie Cantor had the Mad Russian, Jack Benny had Rochester, Dennis Day and Don Wilson, Bob Hope had Jerry Colonna.

Allen, who wrote his own material, had frequent troubles with his sponsors (who wanted to interfere and "improve" his show) and with the network bosses who often made impossible requests of him. At one point they began cutting his show off the air whenever he made a comment they disliked.

The controversy that Allen created was often used to good advantage (as, for example, when he carried on a long make-believe feud with his good friend, Jack Benny, and produced countless new listeners.) Unfortunately, he had difficulty adjusting to television. His last radio program (with Jack Benny as a guest) was on June 26, 1949.

6

Amos 'n' Andy

There is no question that the most famous radio comedy program of all time was *Amos 'n' Andy* (which made its debut on January 12, 1926 as *Sam 'n' Henry*). Freeman Gosden played Amos Jones while Charles Correll played Andy Brown. People were so fascinated by the show (which at one time prohibited studio audiences) that each evening movie theaters had to halt the feature film and turn on the radio at 7:00 p.m. (EST) so that their patrons could follow the latest adventures of the owners of "The Fresh Air Taxi Company." Almost every U.S. citizen knew all about Andy's breach of promise suit by Madame Queen and used catch phrases like Andy's "Buzz me, Miss Blue." President Coolidge stopped work while it was on the air. The show was even quoted in *The Congressional Record.*

With increased interest and concern with civil rights and minority groups, the program later came under attack. However, its creators

6 Fred Allen! — (NBC photo)

7

8

9

7 Fibber McGee and Molly (From *This Way Please*, a 1937 Paramount movie)

8 Bob Hope and friends. (Hope Enterprises Inc.; from *Here Come the Girls,* 1953, Paramount.)

9 Edgar Bergen, Mortimer Snerd and Charlie McCarthy

apparently never dreamed they were creating an unpleasant stereotype. Certainly no one says that Sanford and Son are typical Blacks or that Willie Lump-Lump is a typical white. Probably the greatest concerns of the show's critics were with the portrayal of Blacks by whites in blackface, once a common thing, and the grammatically grotesque creation of a language which is not at all like real Black English.

But Americans, perhaps far more innocent and insensitive than they should have been, passionately followed such madcap antics as The King Fish trying to sell Amos a car. "Put a motor in it and you can drive it right off," he cajoled.

The same type of humor also appeared in later ethnic shows like *Life With Luigi* which also used exaggerated dialects to garner laughs. Another example is *Beulah,* about a Black maid. Her voice was done by Marlin Hurt, a white man. Later, Beulah was played by Hattie McDaniel, Louise Beavers, and still later by Lillian Randolph.

Fibber McGee and Molly

Fibber McGee (Jim Jordan) and wife Molly (Marian Jordan) whose show began on radio in March 1935, had funny routines which were repeated each week—the hall closet full of golf clubs, sleds, baseballs, boxes, and pans which fell out when Fibber opened it, the phone call which always resulted in a brief conversation with Myrt, the operator, and a familiar stream of characters: Doc Gamble (Arthur Q. Bryan), Mr. Whimple (Bill Thompson), The Oldtimer (Cliff Arquette, later known as Charlie Weaver on TV), and Mayor LaTrivia (Gale Gordon). Fibber McGee used at least one tongue-twister per show like "The thicket was not only thick with long thick thorns but every stick was thick with ticks."

The show won listeners who knew what to expect each week at 79 Wistful Vista (the McGee's address). There were spinoffs from the show: *The Great Gildersleeve* and *Beulah.*

Charlie McCarthy

One of the most unlikely radio stars was a ventriloquist's dummy—Charlie McCarthy. After all, the whole point of a ventriloquist's act was to deceive the audience into believing that the dummy and not the ventriloquist was doing the talking. The ventriloquist created this illusion by talking without moving his lips (by substituting certain sounds like V for B). This illusion was obviously lost in radioland.

Edgar Bergen, who began on NBC in 1936, became a hit because of the unusual voices he created for the flip, tuxedoed Charlie and later characters like stupid but good-natured Mortimer Snerd and the old maid, Effie Klinker. Too, a large measure of his program's popularity was owing to his guest stars and humorous insults as when W. C. Fields once said to Charlie, "I'll introduce you to a buzz saw." It was the style of humor to become popular later with Fat Jack Leonard and Don Rickles.

There were other ventriloquists such as Paul Winchell and Jerry Mahoney, Kay Carroll with Tommy, and Shirley Dinsdale with Judy Splinters. None of them ever achieved Charlie's and Bergen's popularity.

Red Skelton and Bob Hope

In the forties, Red Skelton was the darling of the airwaves. He created some characters for his Tuesday night show which he still does on television like Willie Lump-Lump, the bum, and Clem Kadiddlehopper. Some, however, like the Mean Wittle Kid were not suitable for television—Red sounded like a little kid but didn't look like one.

Bob Hope began on radio in 1934 sponsored by Pepsodent. His flip up-to-the-minute monologues sounded as if he had made them up minutes before he came on the air. The show was studded with stars. Much of Hope's humor consisted of a barrage of wisecracks exchanged with Bing Crosby (with whom he made several motion pictures).

While Skelton's humor was slapstick, based on the reaction of his stereotyped characters, Hope's humor was verbal—based on puns and other humorous twists of language.

Abbott and Costello

Most kids in the forties liked the comedy team of Bud Abbott and Lou Costello who made zany movies with the Laurel and Hardy formula—a wise guy and his dumb sidekick. Abbott and Costello's comedy was obvious and slapstick, like that of The Three Stooges; however, Bud and Lou's humor was verbal as well as physical.

After their debut in 1942 on *The Kate Smith Hour,* they got a radio show of their own thanks to such goofy routines as "Who's on First" in which fast repartee with names and words confused Costello but not Abbott or the audience. Shows began with Costello (the fat one) screaming "Hey AA-bbott!"

Other teams who were popular were Moran and Mack, Wheeler and Woolsey, Fishface and Figgsbottle, and Olsen and Johnson.

Comedians and Gimmicks

Every comedian had a gimmick to attract listeners. Judy Canova was a make-believe hill-billy. Most of her wacky humor revolved around being ultra-poor. In one scene, Ma (played by Judy) says, "Pa, the garbage man is coming." Pa replies, "O.K., tell him to leave half a pail."

During the 1940s, Americans went hill-billy crazy. Escaping the tensions of World War II and its restrictions, radio fans found fascination in "Ozark Opry" describing a simple society with few if any rules. Another hill-billy character, Li'l Abner, also appeared on radio in 1939.

My Friend Irma was about another stereotype—a dumb blonde who did goofy things. But off-beat characters got laughs as well. Jimmy Durante rolled 'em on the living room floor with his madcap mispronunciations. Blue collar worker Chester A. Riley (William Bendix) made his fans chuckle whenever he said, "What a revoltin' development this is." One unusual character on his show was Digger O'Dell, an undertaker who always greeted him with, "You're looking fine, Riley, very *natural.*"

Situation Comedies

In addition to these comedy shows and variety shows, there were situation comedies like *Our Miss Brooks,* dating from 1948, about the trials and tribulations of a high school English teacher (Eve Arden) in love with a shy science teacher (played by Jeff Chandler). Richard Crenna played one of the students, Walter Denton.

Beginning November 20, 1929 with Gertrude Berg, *The Goldbergs* was concerned with the life of a poor and lovable Jewish family in New York City. On the opposite side of the coin, there was

10

11

12

10 That pair of zanies, Bud Abbott and Lou Costello

11 *Our Miss Brooks* with Gale Gordon and Eve Arden

12 Jeff Chandler played the bashful science teacher at Madison High on *Our Miss Brooks.* — (Universal Pictures Company, Inc., 1953)

Lum and Abner, a program about two hicks who ran a general store called The Jot 'em Down Store in Pine Ridge, Arkansas. Chester Lauk as Lum Edwards and Norris Goff as Abner Peabody spent their time doing nutty things like getting a lion for their friend's library (they had heard that the New York City Library had lions). Only they got a *real* lion; the New York City Library's were marble statues. Of course, the lion got away and terrified the Pine Ridge townsfolk. Dealing with absurd activities of this sort, the show lasted from 1931 to 1955.

There was *Blondie,* the show that probably made a million youngsters think that all American males were a bit daffy and incompetent and that it took the woman of the family to straighten things out, especially when they heard Dagwood yelling in panic, "Blon - - die!" It wasn't the only show adapted from the comics; there were, for example, Popeye the Sailor, Mr. and Mrs., and others.

Teenagers

A familiar sound on Tuesday nights was Mrs. Aldrich's "Henry! Henry Aldrich!" and Henry's cracked voice answering "Coming, Mother!" The Aldrich family was a situation comedy about teenagers. It may be that the concept of a teenager culture was really unknown before the forties. A child was a child until he was a man. *Andy Hardy* (also a radio show as well as a movie series) is a good example.

Henry Aldrich was the first member of the teenage counterculture. He had troubles with girls, a car, and school like all teens, but most important, he lived in a society of other teenagers (Homer Brown, Kathleen Anderson, Willie Marshall, Agnes Lawson, and Stringbean Kittinger) rather than primarily in his family circle. This change was due to technological advances like the telephone and automobile which widened the cultural peer group.

This development of a new subculture was evident in other shows, notably *A Date With Judy,* which chronicled the teenage tribulations of Judy Foster (played by Ann Gillis) and her boyfriend Oogie Pringle. Two other examples are *Archie Andrews* (based on Bob Montana's comic strip) which began in 1943 on Saturday mornings, and *Harold Teen* (based on Carl Ed's strip) which began August 5, 1941.

Comedy Quiz Shows

It Pays to Be Ignorant was a nutty show with a panel of comedians (Tom Howard, Lulu McConnell, Harry McNaughton, and George Sheldon) who answered simple questions with absurd and improbable answers. For instance, when M.C. Tom Howard read the question, "How much is a two-cent stamp?" their minds went off in different directions. "What color is the stamp?" would be George Sheldon's question and Lulu would demand, in a gravelly voice, "My old man gave me a stamp once on my foot." "Would you repeat the question, Mr. Howard?" Harry McNaughton would chime in. This parody on quiz shows was thought up by Tom Howard's daughter and it went on the air in 1942.

Can You Top This?, a comedy panel show also had many fans who listened to Senator Ford, Harry Hershfield, and Joe Laurie, Jr., swap jokes to score points on a laugh meter.

13 *Lum and Abner* started in 1931 and left the air in 1955. Chester Lauck (Lum Edwards) and Norris Goff (Abner Peabody) have recently revived the show about the Jot 'em Down Store. — (Post Pictures Corp.)

14 Senator Frankenstein Fishface (Elmore Vincent) was a member of a comedy team with Professor Figgsbottle (Don Johnson) in the thirties. The Senator "ran" for governor on the planks of "free feedbags for sea horses and softer cushions for flagpole sitters."

Satire

The fifties brought a more subtle type of humor. The masters were a team named Bob (Elliot) and Ray (Goulding) who satirized radio. They produced shows called *Mr. Trace, Keener Than Most Persons* (a parody of Mr. Keen) and *Mary Backstage, Noble Wife* (a parody of the famous soap opera).

Their humor did not bounce off a studio audience; there was no laughter to punctuate their jokes as Ed Wynn had demanded in the thirties. Thus, at first, their show had a strange quality unlike other radio shows. Their humor was quiet, not raucous. The voices of their many characters were obviously phony. Their women like Mary McGoon or Natalie Attired, for example, each sounded like a male impersonation of an old maid aunt. Yet they were funny.

Why? Good satire is always funny. Whenever something becomes an accepted institution we are eager to poke fun at it. For example, when someone mentions a familiar commercial or uses the same advertising style for bogus product, we laugh. Bob and Ray also poked fun at do-it-yourself shows ("How to tell salt and pepper shakers apart"), sport shows (Steve Bosco, the inebriated sportscaster always ended with "This is Steve Bosco, rounding third and being thrown out at home."), and country and western shows ("Tex Blaisdell" did rope tricks while you listened).

Bob and Ray eventually did a two-hour stage show on Broadway. In the late sixties, they returned to radio with the same format as before. Characters which they created are still part of the show (Wally Baloo, for example). Bob and Ray signed off as always with "Write if you get work and hang by your thumbs."

More Modern Mirth Makers

A product of the seventies was a disc jockey in New York City named Don Imus (*Imus in the Morning*). Although he was (on the surface) one of the people who wake you brightly, he also dealt in a wide swath of crazy characters, daffy sayings and unusual ideas. He was as irreverent as Henry Morgan, who used to ridicule his sponsors. Much of his stuff would have been banned earlier because it often seemed in bad taste. For instance, he parodied religion, poking fun at fundamentalist ministers. He joked heavily about sex and played recordings of outrageously blue conversations on the air. However, like Lenny Bruce, George Carlin, and some of the new breed of comedians, he believed that there was no subject that shouldn't be joked about and nothing that couldn't be said.

Some other groups also appeared to attract a wide audience with their nutty antics: *The Firesign Theatre* and *The Goon Show*. They, too, thrive mostly on satire.

15

16

15 George Jessel did an imaginary telephone conversation with his mother at the end of each comedy routine.

16 The ebullient Jimmy Durante fractured the English language.

1927

The Aldrich Family

Clifford Goldsmith's play (and movie) *What a Life* resulted in this weekly comedy program about a teenager.

Henry Aldrich—Ezra Stone
Sam Aldrich—House Jameson
Alice Aldrich—Lea Penman
Mary Aldrich—Betty Field
Dizzy Stevens—Eddy Bracken
Tommy Bush—Norman Tokar (?)
(Numerous other actors appeared in the cast.)

Director—Day Tuttle
Writer—Clifford Goldsmith
Music—Jack Miller
Announcer—Harry Von Zell

MUSIC.—*Jell-O pudding fanfare.*

MRS. ALDRICH.—*(Calling) Hen-ry! Henry Aldrich."*

HENRY.—*"Coming, mother!"*

MUSIC.—*"This Is It" . . . Fade out behind.*

VON ZELL.—The Aldrich Family! Starring Ezra Stone and written by Clifford Goldsmith, brought to you by Jell-O puddings, those delicious new desserts all America's talking about!

MUSIC.—*Stone opening . . . fade for.*

VON ZELL.—Do you remember when *you* were in your teens? Well, when you listen to Henry Aldrich and his pals, we think you'll sorta be able to detect a little of yourself. For Henry Aldrich is a typical American boy, from a typical American family. Just listen, and see if I'm not right.

MUSIC.—*Out.*

VON ZELL.—As our scene opens we find the Aldriches seated at the dinner table.

MARY.—Have I told you what I'm going to wear, mother?

MRS. ALDRICH.—No, Mary.

MARY.—I'm going to sew maple leaves all over my dress.

MR. ALDRICH.—*Where* are you going to wear a dress like that?

MARY.—To the Halloween party tomorrow night, father. Wherever you look, all you'll see of me is maple leaves.

HENRY.—Maple leaves?

MARY.—Yes.

HENRY.—You're going as a tree?

MARY.—As a tree that's just turned.

MRS. ALDRICH.—Henry, eat your lemon pie.

HENRY.—I don't care for any pie, mother. Father, was there a letter of any kind for me this morning?

MR. ALDRICH.—Nothing that I saw, Henry.

HENRY.—Are you sure you looked carefully?

MR. ALDRICH.—There was a letter, if I remember, for your sister.

MARY.—That was my invitation to the party.

HENRY.—Oh.

MRS. ALDRICH.—And what are you going to wear, Henry?

HENRY.—I, mother? I'm not even going.

MRS. ALDRICH.—Aren't you invited, Henry?

HENRY.—Mother, the reason I'm not going is I don't care to. That's the only reason in the world I'm not going.

MRS. ALDRICH.—*(Impressed)* Well, Henry.

HENRY.—Of course.

MRS. ALDRICH.—Aren't you even going to *taste* your pie, Henry?

HENRY.—I'm not hungry.

MR. ALDRICH.—Let me look at you, Henry.

HENRY.—Father, mother, may I be excused from the table?

MRS. ALDRICH.—If you don't want anything more.

HENRY.—*(Fading)* There's something I want to see in here.

MRS. ALDRICH.—*(Low)* Sam, *what* is the matter with Henry?

MR. ALDRICH.—After all, Alice, just because a boy isn't going to eat a piece of your lemon meringue pie doesn't mean that he is coming down with anything.

MRS. ALDRICH.—Well, I know it isn't the pie. That's the best crust I ever made.

HENRY.—*(Off)* Mary, where did you put that invitation you got?

MARY.—On the living room table, I guess.

MRS. ALDRICH.—*(Low)* Why do you suppose he wants to look at that?

MARY.—*(Calling)* Why do you want to look at it, Henry?

HENRY.—*(Off)* No reason. I was just wondering what kind Kathleen sent out this year.

MRS. ALDRICH.—Mary, *has* Henry received any invitation?

MARY.—Why, I supposed he had.

MR. ALDRICH.—Alice, may I have another piece of that pie, please?

MRS. ALDRICH.—Sam, can't you think of anything but your appetite when something may be wrong with your son?

MR. ALDRICH.—What would you like to have me do?

MRS. ALDRICH.—You can at least call him in and talk to him, dear.

CLEAR CHANNEL STATIONS

272.6—WPG Atlantic City—1100
12:55—Quotations. Municipal Market.
1:00—Luncheon Music.
4:00—World Book Man and Supper Club Entertainers.
4:30—Market Quotations.
4:45—Art Talk. Yvonne DuBayard.
5:00—Alice Reynolds, Pianiste.
5:15—Wm. McIntosh, negro spirituals.
5:30—Organ Recital.
5:45—News.
8:00—Baseball Scores.
8:05—Gospel Hymns.
8:20—Organ Recital.
9:00—Night Club Entertainers.
9:30—Creatore's Band.
10:00—Margaret Keever, Contralto.
10:15—The Subway Boys.
10:30—Pier Orchestra.
11:00—Correct Time.
11:00—Dance Orchestra. Pier Ballroom.
11:30—Orchestra and Novelty Marimba Band.
12:00—Dance Orchestra.
12:30—Casino Dance Orchestra.

348.6—WABC New York—860
6:00—Lone Wolf; Going to Press.
6:30—Duke Ellington's Band.
7:00—United Symphony Orchestra.
7:30—Mac and Lennie.
8:00—Chain Key Station (3 hrs.)
11:00—Hotel Dance Hour.

272.6—WLWL New York—1100
6:00—Orchestra and Vocal.
7:00—K. of C. Hour.

422.3—WOR Newark—710
6:20—Sports; Vocal.
7:00—Hotel Orchestra.
7:30—Slim Figures.
8:00—Lone Star Rangers.
8:30—Wandering Gypsies.
9:00—Character Concert.
10:00—Feature Program.
10:30—To Be Announced.
11:00—News; Dance Hour.

282.8—WBAL Baltimore—1060
7:00—Organ Recital.
7:30—Same as WJZ (2½ hrs.)
10:00—The Liners.
10:30—WJZ Programs (1½ hrs.)

256.3—WCAU Philadelphia—1170
6:20—Scores; Quartet.
7:00—Orchestra, Feature.
8:00—WABC (30 min.); Fur Trappers.
9:00—Hour from WABC.
10:00—Baker Boys; Orchestra.
11:00—Dance & Organ Hour.

305.9—KDKA Pittsburgh—980
6:00—Studio; Scores; Feature.
7:30—Same as WJZ (4½ hrs.); Scores.

260.7—WHAM Rochester—1150
7:30—Hour from WJZ.
8:30—School of Music Program.
9:00—Same as WJZ (2½ hrs.)
11:30—Dance Music (1 hr.)

379.5—WGY Schenectady—790
6:25—Scores; Dinner Music.
7:25—Scores; String Quartet.
8:00—WGY Agriculture Program.
8:30—WEAF Programs (3 hrs.)
11:30—Organ Recital Hour.

302.8—WBZ Springfield—990
6:30—Melodies; News; Scores.
7:00—Melody Boys; Twins.
7:30—Feature; WJZ (1½ hr.)
8:30—Studio; WJZ Programs (2 hrs.)
11:00—Sports; News; Scores.

428.3—WLW Cincinnati—700
7:00—Orchestra; Diners; Scores.

1929

RADIO FACTS!

Variety Shows

Comedy was an important part of variety shows, which also boasted singers, musicians, and dramatic actors. Examples of popular variety shows are: *The Collier Hour, The Chase and Sanborn Hour, The Camel Comedy Caravan, The Comedy Playhouse Laugh Parade, Mirth for Moderns, The Big Show, The Ed Sullivan Show.*

MR. ALDRICH.—Yes, of course. *(Calling)* Henry.

HENRY.—*(Off)* Yes, father.

MR. ALDRICH.—Won't you come here, please?

HENRY.—*(On)* Yes, father.

MR. ALDRICH.—Henry, your mother would like to speak with you.

MRS. ALDRICH.—Sam Aldrich!

HENRY.—You'd like to speak to me, mother?

MRS. ALDRICH.—*(Trying to be very casual)* Tell me, Henry, you've *had* an invitation to go to that party tomorrow night?

HENRY.—An invitation?

MRS. ALDRICH.—Yes.

MR. ALDRICH.—Can't you answer your mother, Henry?

HENRY.—But, father, even if Kathleen begged me, I wouldn't want to go.

MARY.—Why not?

HENRY.—Because, in the first place, I went to her party last year. You certainly don't think I want to go two years in succession, do you?

MARY.—Well, *I* want to go again this year.

HENRY.—But, Mary, you're an entirely different type from me. You're much more easily amused.

MARY.—Why, Henry Aldrich!

HENRY.—But, you are, Mary. The whole thing is so foolish. All they do is dress up in costumes.

MARY.—But they dance, Henry.

HENRY.—Sure. That's another thing. They dance.

MRS. ALDRICH.—All right, dear, all right. But we aren't deaf!

HENRY.—*(In the same pitch)* I'm not shouting, mother. I just want you to understand how I feel. *(Continued)* And besides, Barbara Pearson isn't even going to be at the party.

MARY.—But Betty Walker is.

HENRY.—Betty Walker is?

MARY.—Of course.

HENRY.—Is that right? Well *(fading),* even so, I still wouldn't be interested.

MRS. ALDRICH.—Where are you going, Henry?

HENRY.—*(Off)* Just into the living room.

MRS. ADLRICH.—*(Low)* I'm going in there and talk to him.

MR. ALDRICH.—*(Off slightly, . . . and determined)* May I have a piece of pie, please?

MRS. ALDRICH.—You'll have to eat Henry's.

MR. ALDRICH.—*(Off)* Thank you.

MRS. ALDRICH.—Henry.

HENRY.—What are you coming in here for, mother?

MRS. ALDRICH.—Henry, sit down a minute.

HENRY.—Well?

MRS. ALDRICH.—What *are* you going to do tomorrow night?

HENRY.—Nothing.

MRS. ALDRICH.—Not a thing? Mary is going to be out, and your father and I will be out.

HENRY.—But if Kathleen doesn't want me, I hope you don't think I'd go, do you?

MRS. ALDRICH.—It isn't because she doesn't want you, dear. You must remember, Mary and Kathleen are older than you.

HENRY.—What difference should that make?

MRS. ALDRICH.—Well, there my be times when they would prefer being with boys of their own age. Don't you think there might?

SOUND.—*Door opens off.*

DIZZY.—*(Off)* Hi, Henry! Is it all right to come in without knocking?

HENRY.—Where did you come from, Dizzy?

DIZZY.—*(On)* I just dropped in to see about something. Evening, Mr. Aldrich. Where, where is Mary?

MARY.—*(Approaching)* Here I am.

DIZZY.—Hi! I'll bet you can't guess what I've got in this big box, here.

MARY.—What is it?

DIZZY.—It's my costume for the party. Would you like to have me show you folks a private preview?

MARY.—I don't think you better show it, Dizzy.

HENRY.—I'm not going to the party.

DIZZY.—You're not going, Henry? You're not going!?

MARY.—He wasn't invited, Dizzy.

MRS. ALDRICH.—Mary!

HENRY.—That's not the reason I'm not going.

DIZZY.—But I'm positive you were invited!

HENRY.—You are?

DIZZY.—Sure. At least, I *thought* she told me she was going to invite you.

HENRY.—Father!

MR. ALDRICH.—Yes, Henry.

HENRY.—Are you sure there wasn't any mail for me this morning?

1927

YOUR RADIO TODAY

(Time Is Eastern Standard Throughout)

NEW YORK, July 10 (AP).—A 45-minute program of both sight and sound will be presented on the night of July 21 to inaugurate the new television transmitter. W2XAB, New York, of the CBS chain. The sound part of the program will be sent over a coast-to-coast network, while the pictures will go out only on W2XAB, 2750 kilocycles.

The program will inaugurate a seven-hour daily period on the television station, which is now undergoing a series of tests. Celebrities from the stage, screen and national life are expected to participate.

To determine whether the radio audience desires its dance music in the fast or the slow tempo, a change is to be made in the style of the B. A. Rolfe orchestra period. Each Saturday night for a while on WEAF-NBC, Mr. Rolfe will direct a group of musicians in slow tempo. On Tuesday and Thursday the regular orchestra will play the fast music, with the fans as judges.

Saturday is to bring:

Addresses from London by Premier Ramsay MacDonald, Stanley Baldwin and David Lloyd George, WABC-CBS, 9:30 a. m.

Organ Melodies by Irma Glenn from Chicago, WJZ-NBC, 2 p. m.

Finals of National Professional Tennis tournament, WABC-CBS, 2.

Pacific Feature Hour from San Francisco, Charles Cooper, pianist, WJZ-NBC, 3:15.

Radio play bill, "The Bog Island Murder Trial," WEAF-NBC, 3:30.

Address by Senor Don Salvador De Madariaga, new Spanish ambassador to the United States, WEAF-NBC, 6:45.

Rudy Vallee and his orchestra, WJZ-NBC, 7.

Bernice Claire, Little Jack Little and the Wirges orchestra, WEAF-NBC, 8:30.

Hank Simmons' Show Boat, "The Wrecker's Daughter," WABC-CBS, 9 o'clock.

1931

MR. ALDRICH.—Yes, Henry, I am quite positive.

HENRY.—Well what do you know about that?

DIZZY.—It'll probably come in tomorrow's mail.

MARY.—Henry, I thought you didn't care anything about going.

HENRY.—I don't, Mary. Only don't you think if Kathleen's decent enough to think of me, I shouldn't offend her?

DIZZY.—Gee, the whole town's going to be there.

HENRY.—Who?

DIZZY.—The same gang we had last year.

HENRY.—Yeah? The whole crowd? Is that right?

DIZZY.—Sure.

HENRY.—Father, I know this is unexpected, but could I have enough money to rent a costume?

MR. ALDRICH.—About how much would it come to?

DIZZY.—Well, I'll tell you what I got for $2.

MARY.—Dizzy.

DIZZY.—I don't mind telling you.

MR. ALDRICH.—For $2?

DIZZY.—You can get costumes at any price. Say, Henry, I just had an idea. Why don't I take my costume back, and you and I rent one together?

HENRY.—Just one for the two of us?

DIZZY.—Sure.

HENRY.—What kind?

DIZZY.—They've got a horse there they'll rent for $6.

HENRY.—A horse?

DIZZY.—Sure, we'll go as a horse!

HENRY.—Do you think any girl would want to dance with a horse Dizzy?

MRS. ALDRICH.—Well, whether they would or not, we're not spending any $6 on a costume.

DIZZY.—I'll pay three-fifty, Mrs. Aldrich, if Henry'll let me be the front end.

MRS. ALDRICH.—It seems to me we ought to be able to fix up a very nice costume right here at home.

DIZZY.—In the way of a horse, Mrs. Aldrich.

MRS. ALDRICH.—Perhaps not a horse, exactly.

MARY.—Why don't you go as a ghost, Henry?

HENRY.—I wore a ghost costume once, Mary, and practically suffocated.

MARY.—That shouldn't be so bad.

HENRY.—Did you ever dance with a sheet over your head?

MARY.—I've danced with some boys that might just as well have had.

DIZZY.—You're not referring to me, are you, Mary?

MRS. ALDRICH.—Sam, I know of the very thing.

MR. ALDRICH.—What is it?

MRS. ALDRICH.—It's upstairs in the attic!

HENRY.—In our attic?

MRS. ALDRICH.—Sam, do you remember that costume you wore when we went to that dance just after we were married?

MR. ALDRICH.—I do.

HENRY.—What is it?

MRS. ALDRICH.—George Washington!

HENRY.—George Washington! George Washington! Mother, do you think I want to go to a Halloween party as George Washington?

MRS. ALDRICH.—I certainly don't see why you couldn't.

MARY.—I know where the costume is, mother. *(Fading)* I'll run up and get it.

HENRY.—You needn't get it for me, Mary.

MRS. ALDRICH.—But wait, dear. You haven't seen it.

DIZZY.—Between ourselves, Mrs. Aldrich, I think Henry would look better as a horse.

MR. ALDRICH.—That costume was good enough for me to wear once.

MRS. ALDRICH.—And you took first prize in it.

DIZZY.—Was it a Halloween party, Mr. Aldrich?

MR. ALDRICH.—No, I wouldn't say it was a Halloween party.

HENRY.—In all my life I have never seen George Washington at a Halloween party.

MRS. ALDRICH.—That's the point. No one has ever thought of it before.

HENRY.—Dizzy, would you like to swap with me, and you go as George Washington and take first prize?

MARY.—*(Approaching)* Here it is, Henry. Look, Henry, it's darling!

HENRY.—Blue satin trousers!

MR. ALDRICH.—Of course.

HENRY.—And they're short trousers! What do I wear below them?

MRS. ALDRICH.—You wear your black shoes and a pair of my stockings.

HENRY.—Your stockings, mother!

MRS. ALDRICH.—Who will know the difference?

HENRY.—Well, at least I will? You don't think I could ever hold my head up in your stockings, do you?

BUY DIRECT FROM LABORATORIES!

Amazing New 1936 Super Deluxe

METAL TUBE MIDWEST

18-TUBE

SIX-BAND RADIO

12000 MILE TUNING RANGE

EVERYWHERE, radio enthusiasts are praising this amazingly beautiful, bigger, better, more powerful, super selective 18-tube 6-tuning range radio. It is sold direct to you from Midwest Laboratories at a positive saving of 30% to 50%. (This statement has been verified by a Certified Public Accountant.) Before you buy any radio, write for FREE 40-page 1936 catalog. This super Midwest will out-perform \$200 to \$300 sets on a point-for-point comparison. That is why nationally known orchestra leaders like Fred Waring, Jack Denny, Ted Fio Rito, and others, use Midwest sets to study types of harmony and rhythmic beats followed by leading American and Foreign orchestras.

\$59.50 WITH NEW GIANT THEATRE SONIC SPEAKER (LESS TUBES)

Geo. Olsen Praises Life-Like Tone Realism

Long Island, New York — Midwest out-performs other radios costing almost twice as much. The crystal-clear tone is so life-like that it sounds as though I am in the studios, actually hearing artists performing

30 DAYS FREE TRIAL

TERMS AS LOW AS \$5.00 DOWN

ONLY RADIO COVERING 4½ TO 2400 METERS

PUSH-BUTTON TUNING

ONLY RADIO WITH 6 TUNING RANGES

NEW ACOUSTI-TONE

FULL-SCOPE, HIGH FIDELITY

3 ONE-YEAR GUARANTEES

80 Advanced 1936 Features

Scores of marvelous features, many exclusive, explain Midwest super performance and thrilling world-wide all-wave reception . . . enable Midwest to bring in weak distant foreign stations, with full loud speaker volume, on channels adjacent to locals. Full Scope High Fidelity and brilliant concert tone are achieved, because Midwest enables you to secure complete range of audible frequencies from 30 to 16,000 cycles. Learn about advantages of 6 Tuning Ranges . . . offered for first time: E, A, L, M, H and U. They give tuning ranges not obtainable in other radios at any price! Every type of broadcast from North and South America, Europe, Asia, Africa and Australia is now yours. Send today for money-saving facts.

V-Front Design (Patent Pending)

Deal Direct With Laboratories

No middlemen's profits to pay—you buy at wholesale price direct from laboratories . . . saving 30% to 50%. Increasing costs mean higher prices soon. Take advantage of Midwest's sensational values. As little as \$5.00 down puts a Midwest in your home on 30 days' FREE trial. You're triply protected with: Foreign Reception Guarantee, Parts Guarantee, Money-Back Guarantee!

SAVE UP TO 50%

MAIL COUPON TODAY! FOR *FREE* 30-DAY TRIAL OFFER *and* 40-PAGE FOUR-COLOR *FREE* CATALOG

1936

RADIO FACTS!

More Comedy

Funny stuff was to be heard on *The Candid Microphone* (Allen Funt's predecessor to *The Candid Camera)*, *Here's Morgan* (acerbic wit by a monologist), *Meet Corliss Archer* (one of a vast number of situation comedies), as well as on shows starring a comedian like Milton Berle *(The Milton Berle Show)*. There was even a comic detective show: *McGarry and His Mouse*.

DIZZY.—Look at the coat, Henry.

HENRY.—Sure! It's even got lace on it!

MRS. ALDRICH.—Just put it on, so we can see how you look.

HENRY.—Mother!

MRS. ALDRICH.—Please put it on, dear.

HENRY.—Okay.

MARY.—And here's the wig.

HENRY.—Mary!

MRS. ALDRICH.—Henry!

HENRY.—I'll put it on, but my heart won't be in it.

MRS. ALDRICH.—My goodness, dear. Stand back. Push the wig up off your eyes.

HENRY.—Like this?

MARY.—Not all the way back, Henry.

MRS. ALDRICH.—Dear, you look just exactly the way your father did that night he wore that to the ball.

MR. ALDRICH.—*(Startled)* I looked like that?

DIZZY.—To me he looks quite a little like a horse.

HENRY.—Mother! Couldn't I please rent a costume?

MRS. ALDRICH.—Do you have enough allowance left?

HENRY.—If you'd give me a little extra.

MR. ALDRICH.—I thought we were going to keep within your allowance, Henry.

HENRY.—Say! I know what I can do!

DIZZY.—What?

HENRY.—Never mind! Wait till you see me. Father, could I borrow your hammer and saw tomorrow?

MR. ALDRICH.—What for, Henry?

HENRY.—I'm going to make my costume.

MUSIC.—*Bridge.*

SOUND.—*Hammer pounding on metal.*

MRS. ALDRICH.—*(Off)* Henry! Henry Aldrich!

DIZZY.—*(On)* Henry!

SOUND.—*The pounding stops.*

HENRY.—What'll you have, Dizzy?

DIZZY.—Your mother's calling you.

HENRY.—*(Calling)* You calling me, mother?

MRS. ALDRICH.—Where are you?

HENRY.—Down in the basement.

MRS. ALDRICH.—Doing what?

HENRY.—Making my costume.

MRS. ALDRICH.—What are you making it out of? Sheet iron?

HENRY.—Wait'll you see it! Has the morning mail come yet!

MRS. ALDRICH.—Not yet, dear.

HENRY.—Well, the minute it does, let me know.

MRS. ALDRICH.—*(Fading)* All right.

HENRY.—Dizzy, are you sure Kathleen is sending me an invitation?

DIZZY.—She's sending one to practically everyone else.

HENRY.—Isn't that strange?

SOUND.—*The pounding starts.*

DIZZY.—Henry! Henry! Hey, Henry!

SOUND.—*The pounding stops.*

HENRY.—You speaking to me?

DIZZY.—*Why* won't you tell me what you're making?

HENRY.—Sure, then you'll go and tell everybody else, and all the girls will know who I am.

DIZZY.—You aren't going to go as a locomotive, are you?

HENRY.—A locomotive! *(The pounding begins and stops)* Could you hold this wash boiler up just a minute?

DIZZY.—You're going to put it on, Henry?

HENRY.—Sure. Now wait'll I get my arms through these holes I knocked in it. *(Cloth tears)* Gee did I tear my shirt a little?

DIZZY.—Only about 6 inches.

HENRY.—I've got my arms through.

DIZZY.—Isn't that hole for your neck a little tight?

HENRY.—It feels quite comfortable. Now look, Dizzy, you see those pieces of stovepipe there?

DIZZY.—Don't tell me you're going to put those on, too.

HENRY.—When I raise my right leg, you slip one on.

DIZZY.—*(Grunting as we hear the metal scratching)* When you once get this all on tonight, Henry, how are you going to be moved to the party?

HENRY.—This won't be hard to walk in.

DIZZY.—Noooo!

HENRY.—Now you see those tin cans over there?

DIZZY.—Say, Henry, do you realize this is a dance we're going to?

HENRY.—Put one can on each foot. And don't talk.

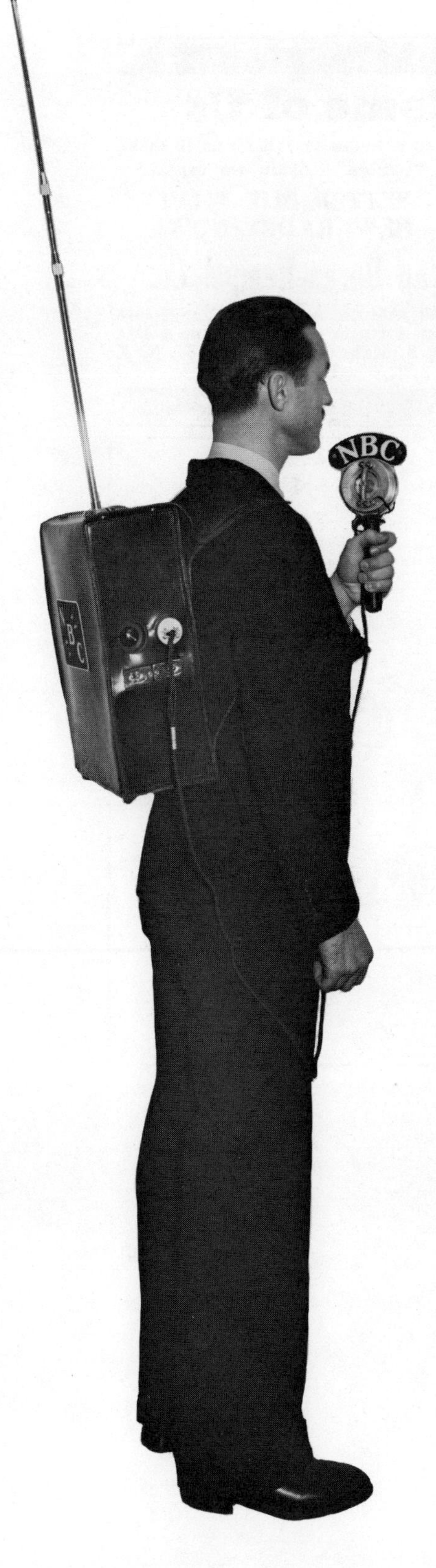

A pack transmitter, September, 1934. (NBC-National Broadcasting Company, Inc., photo)

1942

DIZZY.—Remember, though, if you tip over, you're as good as lost.

HENRY.—Got them on?

DIZZY.—Yes. Now let's see you walk.

SOUND.—*Mild clanking.*

HENRY.—See that, Dizzy. There's practically nothing to it.

DIZZY.—The girls'll have a nice time dancing with you. They just might as well waltz with a meat grinder.

HENRY.—Dizzy, you see that tin funnel right here on the floor?

DIZZY.—Yeah.

HENRY.—Could you please pick it up for me?

DIZZY.—What are you going to do with that?

HENRY.—That's my hat. I'm the Tin Woodman from "The Wizard of Oz."

SOUND.—*Door bell rings off.*

HENRY.—Mother! Mother!

DIZZY.—Are you dying, Henry?

MRS. ALDRICH.—*(Off)* Yes, Henry.

HENRY.—Is that the mailman?

MRS. ALDRICH.—I think it is, dear.

HENRY.—Well, wait. *I'll* be right up.

SOUND.—*Metal clanking as he starts to walk.*

DIZZY.—Henry, you're going to kill yourself.

HENRY.—I'm going to get my invitation. Come on.

SOUND.—*Clanking.*

DIZZY.—You don't think I'm going to walk behind you and risk my life, do you?

SOUND.—*Clanking continues.*

MRS. ALDRICH.—*(Closer)* What in the world is going on there?

HENRY.—Nothing, mother. I'm just coming upstairs.

DIZZY.—It's Frankenstein, Mrs. Aldrich.

SOUND.—*Terrific crash.*

HENRY.—*(Yelling)* Dizzy! Catch me!

MRS. ALDRICH.—*(Screams)* Henry!

SOUND.—*Clanking stops.*

MRS. ALDRICH.—Henry Aldrich, what on earth did you fall into?

HENRY.—I was in it before I fell. It's your wash boiler.

DIZZY.—Stop talking, Henry; and when I lift your head up, hold your legs stiff.

HENRY.—*(Grunting)* Okay.

SOUND.—*Clanking.*

DIZZY.—*(Grunting)* I got you part way up.

HENRY.—Come on now. A little higher.

DIZZY.—If you fall down at the dance, Henry, you needn't count on me. There!

HENRY.—Okay.

SOUND.—*Clanking up the stairs.*

MRS. ALDRICH.—Henry, you can't go to any dance in that outfit.

HENRY.—Mother, it's just a case of getting used to it.

DIZZY.—Of who getting used to it?

HENRY.—Has the mail come?

MRS. ALDRICH.—I suppose it is at the front door.

HENRY.—Well, gee whiz, let me go and get it.

DIZZY.—I'd like to see you tear an envelope open in that outfit.

SOUND.—*Metal scratching on wood.*

MRS. ALDRICH.—Just a minute, Henry, I'll open the door for you. We would still like to use that door.

SOUND.—*Door opens.*

HENRY.—Thank you, mother.

SOUND.—*Clanking continues.*

MRS. ALDRICH.—*Must* you walk on the hardwood floors?

HENRY.—I won't scratch them.

SOUND.—*Metal scratching on wood.*

MARY.—*(Approaching)* Here's the mail, mother.

HENRY.—Could you give me mine, Mary?

MARY.—Well, for goodness sakes, Henry!

DIZZY.—I won't let him hurt you.

HENRY.—Won't you hand me my invitation, Mary?

MARY.—Look at all there is. Here are two letters for mother, one for father.

HENRY.—Never mind those, Mary. I'd like mine.

MARY.—Well, just a minute, Henry. Here's one for me.

DIZZY.—Kathleen's invitations are in yellow envelopes.

MARY.—Here are two more for father. *(Curious)* One of them is from Buffalo, New York.

HENRY.—Mary! Is Buffalo, New York, important at a time like this?

MARY.—There isn't any for you.

HENRY.—I know; but how could a thing like that happen? *(Clanking)*

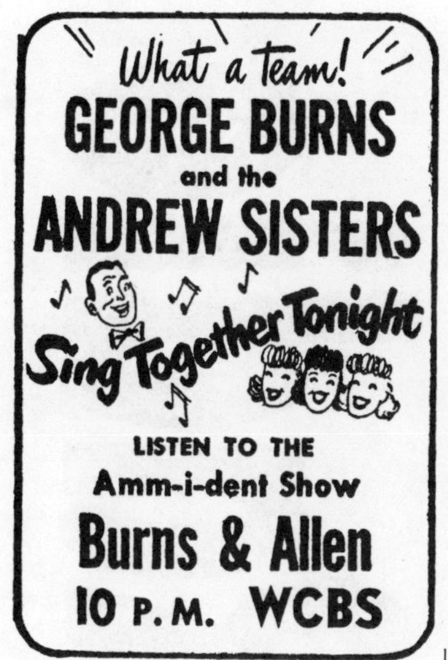

1949

MRS. ALDRICH.—Henry, dear, would you mind not sitting down in that good chair with that wash boiler on? *(Clanking)*

HENRY.—But, gee whiz, all this work for nothing. I thought, Dizzy, you said she was going to invite me.

DIZZY.—Well, don't blame me if she didn't, Henry.

HENRY.—I've got to stay home tonight?

MARY.—Henry, I'm going over to Kathleen's later on, and I'll ask her whether you are invited?

HENRY.—Oh, no you won't, Mary. If she doesn't want to send me an invitation, she needn't.

MRS. ALDRICH.—Mary! If Kathleen does mention the party tonight, I think it would be perfectly all right to at least give her a chance to ask whether Henry is coming.

MARY.—Of course it would.

DIZZY.—You're marring the wallpaper, Henry.

HENRY.—Well, can you stand me up, please?

MARY.—Shall I speak to Kathleen, Henry?

HENRY.—Don't ask her outright, Mary. Just say it's too bad Henry hasn't any place to go tonight. Sort of subtle, see?

MUSIC.—*Play off.*

(Commercial)

MUSIC.—*Sneaks in during last few sentences . . . then up to a finish.*

MRS. ALDRICH.—*(Calling)* Mary!

MARY.—*(Off)* Yes, mother!

MRS. ALDRICH.—Will you ask Henry to come down here to the living room, please?

MARY.—Yes, mother. I'll call him.

MR. ALDRICH.—It seems rather late, though, to be breaking news like this to him, Alice.

MRS. ALDRICH.—Sam, do you think we really should go to that bridge party this evening?

MR. ALDRICH.—But we accepted the invitation weeks ago.

MRS. ALDRICH.— But, Sam, we ought to stay home with Henry. It's Halloween.

SOUND.—*Slow clanking off.*

HENRY.—*(Approaching)* Where are you, father?

MR. ALDRICH.—Right here, Henry.

HENRY.—*(On)* Just a minute while I make this turn.

MRS. ALDRICH.—Henry, your father wants to speak with you.

HENRY.—With me, father? With me?

MRS. ALDRICH.—I think I'll go upstairs.

MR. ALDRICH.—Henry, your mother has asked me to give you a rather unpleasant message.

HENRY.—Well?

MR. ALDRICH.—Mary was over at Kathleen's a few minutes ago.

HENRY.—Well?

MR. ALDRICH.—I'm afraid, son, you're not invited.

HENRY.—Kathleen said I wasn't?

MR. ALDRICH.—All I know is Mary assumed from Kathleen's conversation you were not.

HENRY.—But, gee whiz, I must be. I'm practically the only one in the whole crowd that isn't.

MR. ALDRICH.—*(Kindly)* Don't you suppose there will be other Halloweens?

HENRY.—I certainly hope there won't be any more like this.

MR. ALDRICH.—Henry, there are going to be a great many times in your life when things won't go exactly the way you'd like to have them. And when they do, there's only one way to take them.

HENRY.—But, dad, this is so trivial. Kathleen just forgot to ask me.

MR. ALDRICH.—Whether she forgot or not, the gentlemanly thing to do is to stay away from any party to which you are not invited.

MARY.—*(Approaching)* Father, where's mother?

MR. ALDRICH.—Upstairs, I think, Mary.

MARY.—*(On)* Henry, do you think I look like a maple tree?

HENRY.—*(Holding back the tears)* Sure. *(Clanking)*

MARY.—Where are you going?

HENRY.—*(Fading)* Up to my room. Where do you think? And take this off. *(Clanking)*

MR. ALDRICH.—Why are you putting on your coat this early, Mary?

MARY.—I'm taking some cakes over to Kathleen's, father. And when Tommy Bush comes back, tell him his costume is right here in the hall.

MR. ALDRICH.—He's going to put it on here?

MARY.—It's so much nearer to Kathleen's if he does.

SOUND.—*Door opens.*

MRS. ALDRICH.—*(Off)* Don't stay out too late, Mary.

MARY.—I won't. Good-by.

SOUND.—*Door closes.*

MRS. ALDRICH.—Get your things on, Sam.

MR. ALDRICH.—*(Calling)* Henry! Henry!

HENRY.—*(Off)* Yes, father.

MR. ALDRICH.—Come down here a minute.

1941

ZENITH "TRANSOCEANIC"

124.40

Consistently fine performance on broadcast and five international shortwave bands. Removable Wavemagnet. Pop-up Waverod for shortwave. Radi-organ tone control. Black stag luggage case, Gold color trim. Carrying handle. AC-DC or batteries. Batteries extra.

POWER PACKED RADIO VALUES

1948

HENRY.—*(Approaching)* You changed your mind? I can go anyhow?

MRS. ALDRICH.—No, dear. Will you try to understand if your father and I go out to an engagement that we promised to go to a long time ago?

HENRY.—I'll be here all alone?

MRS. ALDRICH.—You mean to say a boy as old as you minds being alone?

HENRY.—No, I don't mind, mother. But I'm going to be *all* alone?

MRS. ALDRICH.—If we didn't have to go, we wouldn't think of leaving you.

MR. ALDRICH.—If anyone comes to the front door, Henry, I'd be very careful whom I let in.

HENRY.—You think anything might happen?

MR. ALDRICH.—No, not a thing. But on Halloween I'd simply be careful. Put the chain on the door.

HENRY.—Sure.

MRS. ALDRICH.—Good night, dear.

HENRY.—Good night.

MR. ALDRICH.—Good night, son.

SOUND.—*Door opens.*

MRS. ALDRICH.—Don't sit up too late.

HENRY.—I won't.

MRS. ALDRICH—*(Off)* Good night.

SOUND.—*Door closes . . . Telephone rings . . . it rings again . . . slow clanking*

HENRY.—I'm coming to the phone. *(Clanking . . . receiver lifts)* Hello Who? Mary isn't in right now. Mrs. Aldrich isn't in either. No, ma'am. There's nobody here but me. Is that you Aunt Harriet? Gee, I'm certainly glad to hear your voice. No I don't mind being alone. What is there to be afraid of? *(Door bell rings)* I wonder what that could be? Our front doorbell rang. *(Doorbell rings . . . voice low)* Hold the line, while I look out the window and see who it is. *(Pause . . . clanking . . . doorbell rings . . . pause . . . clanking . . . voice low)* Aunt Harriet, are you still there? I think it's a man in a brown coat. Yeah, he's got a hat on. *(Doorbell rings)* I'm not afraid. Gee whiz, what is there to be afraid about? You wouldn't like to get in your car and come over, would you?

SOUND.—*Doorbell rings . . . pounding on door.*

DIZZY.—*(Off)* Hey, Henry! Can't you let a guy in here?

HENRY.—*(Coming to life)* Aunt Harriet, I've got to let Dizzy in! Good-by! *(Receiver hangs up . . . clanking . . . pounding on door)* Is that you I see through the window, Dizzy?

DIZZY.—*(Off)* Who do you think it is?

SOUND.—*Door opens.*

HENRY.—Hi!

DIZZY.—Hi. Come on, Henry. Let's start for the party.

HENRY.—I'm not going.

DIZZY.—You're not going?

HENRY.—Kathleen didn't invite me.

DIZZY.—All right, what if she didn't? Come along anyhow.

HENRY.—No.

DIZZY.—Who'll know the difference?

HENRY.—Mary will be there.

DIZZY.—What's that package there on the floor?

HENRY.—How should I know what it is?

DIZZY.—It's from the Century Costume Company. That's where I got my outfit.

HENRY.—It's probably something Mary was going to use.

DIZZY.—I wonder what it is? Do you mind if I open it just a little?

HENRY.—I don't care whether you open it.

SOUND.—*A string breaks . . . paper rattles.*

DIZZY.—Henry! Look!

HENRY.—What about it?

DIZZY.—Look at it! It's the horse they wanted to rent me!

HENRY.—*(Interested)* Yeah?

DIZZY.—Henry, why don't we use this?

HENRY.—It isn't ours.

DIZZY.—Is Mary going to wear it?

HENRY.—Of course she isn't.

DIZZY.—Is your father going to?

HENRY.—*(Pleading)* Listen, Dizzy, my father said it wouldn't be right for me to go any place I'm not invited.

DIZZY.—Who's going to know you're even there? I'll be in the front legs, and you'll be in the back. Kathleen and Mary won't even see you.

HENRY.—You don't think so?

DIZZY.—How could they? Here, put on the pants for the hind legs.

HENRY.—Okay.

DIZZY.—I'll be getting into the front legs. Here, throw this blanket over your head. *(At this point the conversation of both boys becomes slightly muffled)* I'll put my head up inside the horse.

HENRY.—Okay.

Wavelengths of New York Stations at a Glance

WMCA	570	WNYC	830	WINS	1010	WFAS	1240	WHOM	1490
WEAF	660	WABC	880	WHN	1050	WOV	1280	WBYN	1430
WOR	710	WPAT	930	WNEW	1130	WEVD	1330	WQXR	1560
WJZ	770	WAAT	970	WLIB	1190	WBNX	1380	WWRL	1600

FRIDAY, SEPTEMBER 17, 1948

Radio Timetable

WAAT—970k
WBBR—1330k
WBNX—1380k

WEVD
WFAS-
WHLI-

FRI. A.M.	WMCA 570-k	WNBC 660-k	WOR—MUTUAL 710-k	WJZ-77
7 00 15 30 45	News; Songs Morning Herald Classified Col. Unity Views	News; Bob Smith Show News—McCarthy Bob Smith Show	News—M. Elliott The Musical Clock	News—G Kiernan' Corne (7:55) N
8 00 15 30 45	News; Joe Frankin's Antique Record Shop	News; Bob Smith Show Tex & Jinx interviews	News—Robinson Dorothy and Dick	News—A Ed and Fitzge (8:55) G
9 00 15 30 45	News; The Duke Ellington Show	News—Roberts Ivan Sanderson Norman Brokenshire	News—Hennessy Passing Parade The McCanns at Home	Breakfas Variet with D McNeil
10 00 15 30 45	News; The Ted Steele Show	Fred Warmg Show Road of Life Joyce Jordan	News—Gladstone Martha Deane reviews and Interviews	My True Story Betty Cr Listening
11 00 15 30 45	News; Live and Recorded Music Cecil Brown	Nora Drake Love and Learn Jack Berch Lora Lawton	News—Robinson Tello-Test quiz Heart's Desire	Breakfas Hollyw Ted Ma Walter H
12 00 15 30 45	News; Mr. and Mrs. Music —records and interviews	News—McCarthy Local News Norman Brokenshire	K Smith Speaks K. Smith Sings News—Gladstone The Answer Man	Welcome Trave' News; N Craig
1 00 15 30 45	News; Mr. and Mrs. Music —records and interviews	Mary Margaret McBride Show	Luncheon at Sardi's John Gambling V. H. Lindlahr	News—B Nancy C People c Things
2 00 15 30 45	News; Ted Steele Show (2:25) Baseball: Giants	Double or Nothing Today's Children Light of World	Queen for a Day On Your Mark Melodies	Maggi M & H. S Bride an Groom
3 00 15 30 45	vs. Reds at Polo Grounds. Also heard on WOR-FM	Life Beautiful Ma Perkins Pepper Young Right Happiness	Movie Matinee Daily Dilemmas	Ladies Be Sec Second Honey
4 00 15 30 45	Ted Steele Show	Backstage Wife Stella Dallas Lorenzo Jones Widder Brown	The Barbara Welles Show Tiny Ruffner— Ladies' Man	Listen to This Treasury Nelson
5 00 15 30 45	News; Mr. and Mrs. Music —records and interviews	A Girl Marries Portia Faces Life Just Plain Bill 1st Page Farrell	Woody & Virginia Superman Adventures Tom Mix	Challeng the Yu Jack Ar Adven
6 00 15 30 45	News; Story; Music; Movies Race Results S Ellis, sports	News—Banghart Bill Stern Wayne Howell Three Star Extra	News—Lyle Van Bob Elson News Stan Lomax	News; S Ethel & Hill; Pres Allen Pr
7 00 15 30 45	News; Music Styled for Strings— Studio Music	Sammy Kaye World News Melody Riders H. V. Kaltenborn	Fulton Lewis The Answer Man Henry J. Taylor Inside of Sports	Headline Elmer Do The Lone Ranger
8 00 15 30 45	News; Big Time Echoes News—Walsh E. Larkin—piano	Band of America Who Said That?—quiz	Ice Follies of 1949 Leave It to Girls (8:55) Billy Rose	The Fat myster This is Y FBI—dr
9 00 15 30 45	Concert on the Mall Inside New York	University Theater— The Red Skelton Show	Gabriel Heatter Frank Leahy Share the Wealth (9:55) News	Break the Bo The Sher (9:55) Sp
10 00 15 30 45	Songs by Perry Como News UN Today	Life of Riley Sports Newsreel War Memorial	Meet the Press Symponette— M. Piastro	Cavalcad of Spo Your Am Sports
11 00 15 30 45	News; Music Unity Views South of Border Music	News—Banghart Dick Dudley Pastels In Rhythm	Vandeventer News; Weather Dance Music	News J. Hasel Dance Music
After Mid	Music and News All Night	News: 12 and 12:55. Off till 5:30	Music and News all Night	News and 12 Off til

1949

DIZZY.—Now put your hands on my shoulders. Can you see?

HENRY.—Did you ever try to see in pitch dark? Let's go.

DIZZY.—How are we going to open the front door?

HENRY.—That's for you to figure out. Can you find it?

DIZZY.—I think so. No. Wait a minute, Henry, these are the hall stairs we're starting up.

HENRY.—This isn't going to lead to any good, Dizzy.

SOUND.—*Door opens.*

DIZZY.—I've got the front door now. You close it as we go out.

HENRY.—Sure. I hope I closed it.

DIZZY.—Be careful of the front steps, Henry. *(We hear them going down the steps)* Are you down yet?

HENRY.—How would I know?

DIZZY.—Come on, now, let's run.

HENRY.—You're crazy!

DIZZY.—Come on! Get up, there! *(There's a thud)* Ouch! Henry! I think I broke my neck.

HENRY.—What's the matter?

DIZZY.—I think we hit a tree!

MUSIC.—*Bridge . . . segue to dance music from a piano . . . voices . . . Halloween horns and general confusion . . . the voices of Dizzy and Henry are still muffled when the speak.*

HENRY.—Dizzy! Dizzy!

DIZZY.—Where are you Henry?

HENRY.—I'm right where I've been for the last 2 hours. Inside this horse.

DIZZY.—You ever have a better time, Henry? We've danced with every girl here.

HENRY.—What fun do you think I've had?

DIZZY.—Four times we've danced with your sister.

HENRY.—She probably thinks we're somebody else.

DIZZY.—Keep quiet, Henry. *(Falsetto)* How do you do, Miss. Would you like this dance? Oh, pardon me!

HENRY.—What's the matter?

DIZZY.—That's a chair.

MARY.—Hello, Tommy.

TOMMY.—*(Sulking)* Hello, Mary. What I want to know, Mary, is who's inside that horse.

MARY.—Why?

TOMMY.—There are several reasons. And if it's the one I think it is . . .

HENRY.—*(Low)* Come on, Dizzy. Let's get away from here.

DIZZY.—Maybe we better.

HENRY.—Let's go over and try to get some punch.

MUSIC.—*Begins again.*

DIZZY.—I can't drink any more.

HENRY.—Maybe you can't, but I haven't had any yet. Come on.

DIZZY.—How are you going to drink under there?

HENRY.—Can't you pass it under your legs.

DIZZY.—Okay. Come on. *(A terrific crash of glass . . . people scream)* Let's go.

HENRY.—What was that.

DIZZY.—That was the punch.

HENRY.—Tell me, Dizzy, have you seen Betty Walker?

DIZZY.—Sure. We just finished dancing with her, you dope.

HENRY.—With my girl? Are the decorations nice?

DIZZY.—They're swell.

HENRY.—I think I would have been a lot better off if I'd stayed home.

DIZZY.—*(Falsetto)* How do you do, Mary Aldrich. Would you like to dance once more?

MARY.—Won't you tell me who you are?

DIZZY.—I'm the Prince Charming that was turned into a dragon. Come on, Mary.

HENRY.—Dizzy, this is no time to tango.

DIZZY.—I like to tango.

MARY.—What was that?

DIZZY.—Nothing. I was just talking to myself.

TOMMY.—Wait a minute there! Who's inside that horse?

SOUND.—*Thud.*

HENRY.—Who hit me?

TOMMY.—I did.

MARY.—Tommy Bush!

SOUND.—*Thud.*

DIZZY.—Cut it out. Do you want to choke me to death?

TOMMY.—Let me pull that thing off your head!

MARY.—Henry Aldrich! Where did you come from?

HENRY.—I?

TOMMY.—I *thought* this was where my costume went.

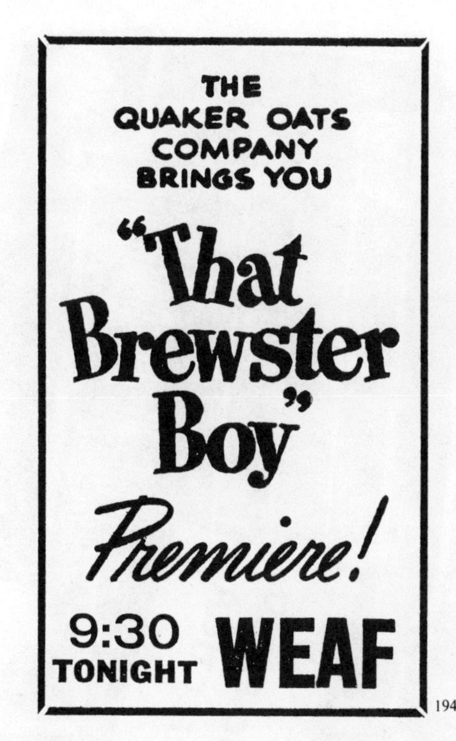

1941

1948

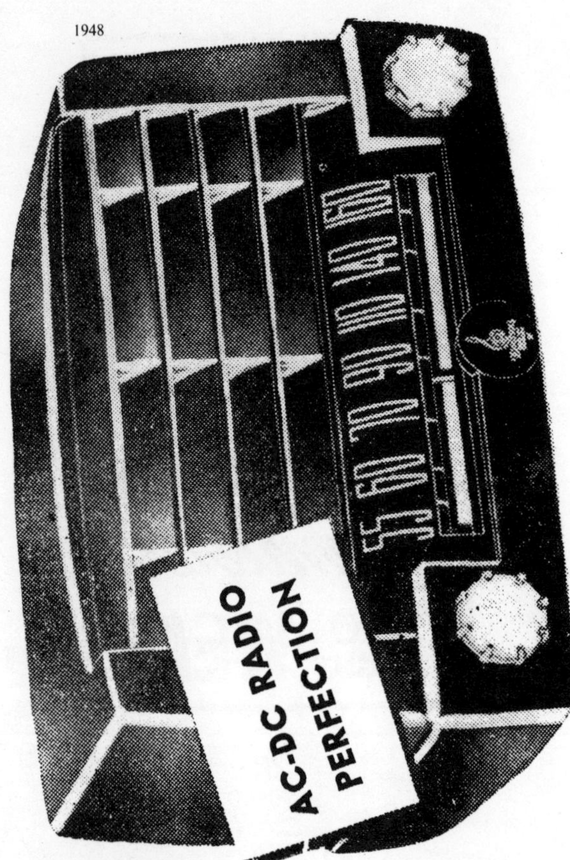

Emerson
Radio

Mary.—Henry, do you realize father has telephoned here twice for you?

Henry.—He wants me?

Tommy.—And you can pay for this costume!

Dizzy.—Where are you going, Henry? I think I'd better be leaving.

Henry.—I'm going home.

Music.—*Bridge.*

Mr. Aldrich.—Henry, I don't believe I've ever been quite so disappointed.

Henry.—But, father. I don't think Kathleen even knew I was there.

Mr. Aldrich.—That isn't the point, Henry. I left you here on your honor. Not only did you take a costume that didn't belong to you, but when your mother and I came home, we found the front door wide open.

Henry.—*Wide* open?

Mr. Aldrich.—Well, Henry, what are we going to do about it?

Henry.—What would you suggest, father?

Mr. Aldrich.—I think it would be much better for you to do the suggesting.

Henry.—I have to tell you what my own punishment is to be?

Mr. Aldrich.—You do. And I want you to tell me quickly.

Henry.—Well, would you—would you be willing not to give me any allowance for the next four weeks.

Mr. Aldrich.—I think that is an excellent suggestion.

Henry.—All right. No allowance for the next four weeks.

Mr. Aldrich.—And what else would you suggest.

Henry.—What else?

Mr. Aldrich.—That is what I said.

Henry.—Well, suppose I go to bed an hour early every night next week?

Mr. Aldrich.—That is agreeable with you?

Henry.—Yes, sir. I realize I've got to learn sometime, and I might as well get it over.

Mr. Aldrich.—Would you like to suggest anything more?

Henry.—Anyting more? Well, I'm pretty sure that's enough to make me *think* the next time.

Mr. Aldrich.—And you have nothing else to suggest?

Henry.—Supposing I help mother with the dishes all next week, too?

Mr. Aldrich.—You've never had a punishment as severe as this, have you, Henry?

Henry.—I'll say I haven't.

MR. ALDRICH.—You're quite sure you feel it is only fair.

HENRY.—Yes, sir.

MRS. ALDRICH.—*(Approaching)* Sam.

MR. ALDRICH.—Yes, Alice.

MRS. ALDRICH.—Are you and Henry through?

MR. ALDRICH.—We are. Supposing you march up to bed, son.

HENRY.—Good night, mother.

MRS. ALDRICH.—Good night, dear.

HENRY.—And I want you to know, father, I feel pretty much ashamed to think I went to the party.

MRS. ALDRICH.—I'm glad to hear you say that, dear.

MR. ALDRICH.—No allowance next month, to bed an hour early every night next week, and you're to help your mother with the dishes every night.

HENRY.—Yes, sir. And if you can think of anything else you'd like to have me do, I'd be glad to do that, too.

MR. ALDRICH.—That will be all.

HENRY.—*(Off)* Good night, father.

MRS. ALDRICH.—*(Low)* Sam, no boy ever lived that didn't disobey his parents now and then.

HENRY.—*(Off)* Father.

MR. ALDRICH.—Yes, Henry.

HENRY.—I almost forgot to tell you. Aunt Harriet phoned just after you left.

MRS. ALDRICH.—What did she want, dear?

HENRY.—She said she wrote you a letter a couple days ago and wondered why you hadn't answered it.

MRS. ALDRICH.—Thank you, Henry.

MR. ALDRICH.—I remember getting it yesterday morning. I put it here on the living room table.

MRS. ALDRICH.—It isn't there now, dear.

SOUND.—*Drawer opens.*

MR. ALDRICH.—Here it is in this table drawer with the rest of my mail.

MRS. ALDRICH.—What does she say?

MR. ALDRICH.—Well, well I wonder . . .

MRS. ALDRICH.—Is it some bad news?

MR. ALDRICH.—I'm not even looking at her letter. I'm looking at this one addressed to Henry.

MRS. ALDRICH.—To Henry? Sam Aldrich, that's Henry's invitation!

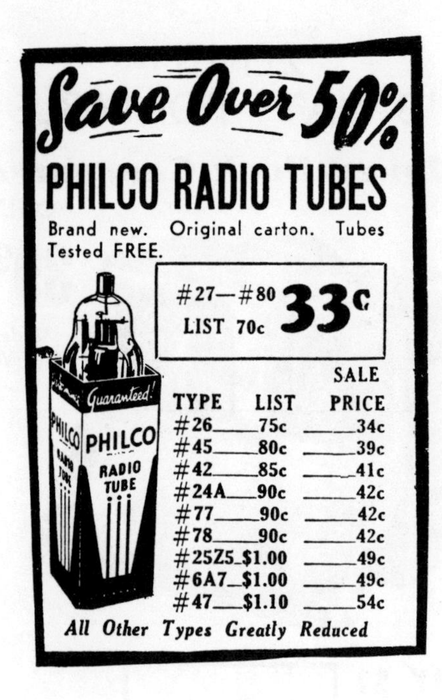

MR. ALDRICH.—How did it get in with my mail?

MRS. ALDRICH.—You got the mail yesterday morning yourself, Sam. I remember distinctly your getting it.

MR. ALDRICH.—Yes, yes. Henry, Henry!

HENRY.—*(Off)* Yes, father?

MR. ALDRICH.—Could you come down here a minute, please?

HENRY.—I've only got half of my pajamas on, father.

MRS. ALDRICH.—Please come down here at once! Sam Aldrich I could cry.

HENRY.—*(Approaching)* Did you think of something else you want me to do, father?

MR. ALDRICH.—Henry, I just found something of yours in this top drawer.

HENRY.—You did?

MR. ALDRICH.—Supposing during next month, we double your allowance.

HENRY.—You double it?

MR. ALDRICH.—And if you would like to, you may stay up fifteen minutes later than usual each evening next week.

HENRY.—Are you kiding me?

MR. ALDRICH.—And, Alice, during the same period, I'll help with the dishes.

HENRY.—Mother, is father out of his head?

MUSIC.—*Aldrich play off . . . segue to fast tune . . . fade for*

(Commercial)

MUSIC.—*Up to finish.*

HENRY.—Mother.

MRS. ALDRICH.—Yes, Henry.

HENRY.—Would you like to have me go over to the store and get some soap?

MRS. ALDRICH.—Do we need any?

HENRY.—That isn't the point. I'm going to take up music.

MRS. ALDRICH.—I beg your pardon?

HENRY.—There's an advertisement here that says if you save 3,000 soap wrappers, you get a silver cornet free.

VON ZELL.—*(Chuckling)* Be sure to listen next week at this same time for more adventures of Henry Aldrich. The Aldrich Family, starring Ezra Stone, is written by Clifford Goldsmith. Original music composed and conducted by Jack Miller! Harry Von Zell speaking and wishing you good night, for those delicious new desserts all America's talking about Jell-O puddings!

MUSIC.—*Jell-O pudding fanfare.*

VON ZELL.—Erza Stone is now appearing at the Biltmore Theatre in the George Abbott Farce "See My Lawyer."

1929

AFTERNOON and EVENING RADIO FEATURES

THURSDAY, JULY 18
(By The Associated Press)

Programs in Eastern Daylight Saving time. All time is P. M. unless otherwise indicated. Wavelengths on left of call letters, kilocycles on right. Clear channel stations and chain programs with list of associated stations in detail.

348.6—WABC New York—860

8:00—Lopez Orchestra—Also WFAN WEAN WKBW WCAO WJAS WLBW.
8:30—United States Marine Band Concert—Also WNAC WEAN WFBL WKBW WCAO WJAS WLBW WMAL.
9:00—Detective Mysteries—Also WCAU WNAC WEAN WFBL WKBW WCAO WJAS WLBW WMAL WSPD WHK WADC WGHP.
9:30—Buffalo Symphony Orch.—Also WCAU WNAC WEAN WFBL WKBW WCAO WJAS WADC WGHP WSPD WHK WLBW WMAL.
10:00—Voice of Columbia—Also WNAC WEAN WFBL WADC WCAO WKRC WHK WGHP WLBW WFAN WJAS WSPD WMAL WKBW.

454.3—WEAF New York—660

6:00—Black & Gold Room Orchestra—Also WFI WRC WCAE WWJ.
6:55—Scores—WEAF; Midweek Hymn Sing—Also WCSH WRC WFJC WJAR.
7:30—Comfort Music—Also WEEI WTIC WJAR WTAG WCSH.
8:00—Buck and Wing—Also WTAG WFI WCAE WRC WTIC WWJ.
8:30—Broadway Lights—Also WJAR WFI WRC WGY WGR WTAM WWJ WSM WAPI.
9:00—Singers, Male Quartet and Violins—Also WEEI WTIC WJAR WTAG WCSH WFI WRC WGY WGR WCAE WTAM WWJ WSAI KYW WFJC.
9:30—Historic Trials, "Joan of Arc"—Also WJAR WTAG WRC WCAE WWJ WGY.
10:00—Old Counselor's Reception, with Andy Sannella's Orchestra—Also WTIC WJAR WTAG WCSH WFI WRC WGY WGR WCAE WWJ WSAI KYW WAPI WHAS WSM WSB WBT WRVA WEEI.
10:30—Concert Bureau, all male entertainment—Also WFI WRC WGY WGR WSAI WRVA WIOD WFJC WTAG WCAE WWJ WSM.
11:30—Jack Albin and His Hotel Dance Orchestra—Also WWJ.
12:00—Dave Bernie's Hotel Dance Orchestra (one hour)—Also WRC WSM.

394.5—WJZ New York—760

6:00—Old Man Sunshine, Bob Pierce's Stories; Scores—WJZ.
6:30—Ben Pollack's Hotel Dance Orchestra—Also WJR KYW.
7:00—Talk—"Street Traffic"—WJZ.
7:30—To Be Announced—WJZ.
8:00—Beauty Serenade with Male Trio and Orchestra—Also WBZ WHAM KDKA WJR KYW WBAL WLW.
8:30—The Ghost Hour—Also KDKA WJR.
9:00—Gus Haenschen Orchestra—Also WBZ WBAL WHAM KDKA WJR WLW WAPI WHAS WSM WSB WBT WRVA WPTF.
9:30—Rosario Bourdon's Concert Orchestra—Also WBZ WBAL WHAM KDKA WJR WLW WHAS WSM WBT WJAX WRVA WSB.
10:00—Dance Orchestra—Also WBZ WHAM WJR WBAL KDKA WGN.
10:30—Around the World, Soprano and Mixed Quartet—Also WBZ WHAM KDKA WJR KYW WBAL WLW WHAS WSM WSB WAPI.
11:00—Slumber Music Hour, String Ensemble—Also WRC KDKA WBAL WHAM.

Halo, everybody, Halo,
Halo is the shampoo that
glorifies your hair.
So, Halo, everybody, Halo.

HELP!

Chapter 3

Yipes! You enjoy being scared, don't you? C'mon now—admit it. The more chills—the more thrills. If you see a monster in the movies (or on TV), you know it's only a lizard or a mechanical monster, enlarged by trick photography. But on radio, the monster was real—you saw it in your imagination. Thus, it was more frightening.

The First Radio Drama

On August 3, 1922, the WGY Players (Schenectady, New York) presented the first radio drama, *The Wolf.* Every Friday thereafter they broadcasted a two-and-a-half-hour-long play. Later, the plays were cut to ninety minutes.

Other stations began to broadcast drama. On November 9, 1922, WLW did its first play and again on April 3, 1923—*When Love Awakens,* an original radio drama by Fred Smith (manager of WLW). The WGY Players performed the first network drama in April, 1924. Radio shows called *Biblical Dramas, True Story, Real Folks, Main Street* appeared in 1928 and 1929. On September 24, 1933 *Roses and Drums* reached the airwaves via WABC (New York). The show was broadcast from a costumed cast on a Broadway stage.

Drama Technique

Several tricks are used to keep the listener aware of who is speaking. First, actors call each other by name (especially in the first lines of a scene). Sometimes a person in the scene doesn't speak until later; the scene has to be set so that listeners know he is there. Actors may describe the person they are speaking about (" Paul, my brother-in-law") or identify themselves ("I'm your long lost husband.").

The listener gets information only through a narrator, dialogue, sound effects and music. If most of the information comes from the narrator, the play seems dull, wooden, and stodgy. It is much more interesting when the play has a great deal of dialogue. Sound effects add another dimension if they are not overdone. Too, they must be clear so that they do not have to be explained. That is, a gunshot should not sound like something else—say, a drawer slamming. The sounds do not need to be done for everything nor do they need to occupy the same time as in real life (for instance, walking up ten flights of stairs or hammering together a wooden bench). Radio shows need to be fast paced. "Well, I'm going to go. Goodbye," says someone and immediately there is the sound of a door closing.

Music adds to the radio play by affecting the listener's subconscious. Some musicians who wrote scores for radio dramas went on to Hollywood (Bernard Herrmann is one example). Sad, but true, few radio composers achieved fame.

How Long Should a Radio Drama Be?

In the beginning, radio shows varied in length, but eventually programs ran in multiples of five minutes—almost always in fifteen or thirty minute segments. The fifteen-minute time slots were filled by the serials (programs which were continued from day-to-day). The half-hour spots were often devoted to drama which was similar to a complete short story. There was little room for character development; usually, the story had a surprise or "O. Henry" ending. A full hour was used for more complex and complete shows like the movie adaptations presented on *Lux Radio Theatre.* This meant that radio broadcasts, like sonnets, had a precise length. On the air, if a program ran short, the extra time was filled by musical "bridges."

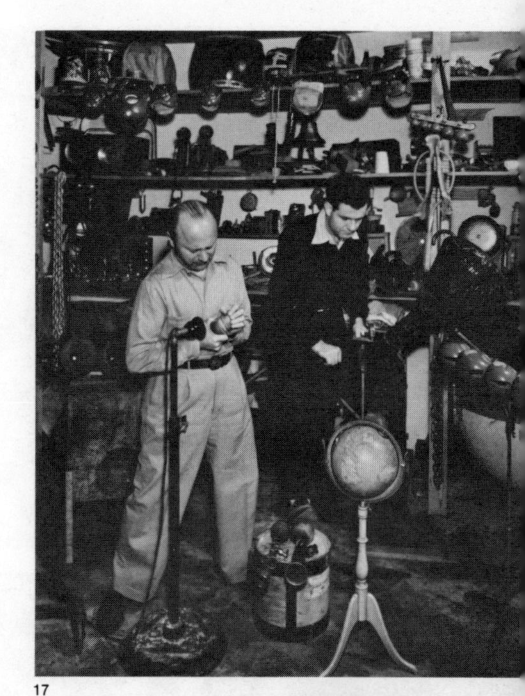

17

18

17 Radio sound effects men created the sound of fire crackling by crumpling cellophane, the sound of walking on snow by squeezing a box of cornstarch, or the sound of thunder by pounding sheets of metal.

18 Luis Van Rooten *(Crimes of Carelessness)*

20

21

As radio writer Rome Cowgill points out, short scenes make a radio show seem fast and lively; long scenes give depth.

According to Cowgill, four characters are the usual maximum for a radio play, two men and two women. A total of eight or more characters resulted from having each of the actors "double" (change his voice slightly). Too many voices are confusing because you tend to forget who's who.

Gangbusters

Phillips H. Lord was the first great adventure program writer. He began producing a radio program called *G-Men* in the thirties but later changed the name of the program to *Gangbusters.* The program presented law enforcement officers and their cases, the more sensational the better. Unfortunately, in real life, detective work often is not very glamourous, but consists of making phone calls and asking tedious questions. Most crimes are solved by tips or by luck unlike the razzle-dazzle effects of the program.

Paying writers $200 a script, Lord based *Gangbusters* on real crimes dealing with counterfeiters, con-artists, and bank robbers.

The program had a policeman narrate the case by proxy, which meant that a professional actor read his part. The show used lots of sound effects—whistles, sirens, and gunshots. It gave us the expression "coming on like *Gangbusters*"·meaning anything done very aggressively. *Gangbusters* always proved that crime doesn't pay.

Detective Shows

Every detective series on the air had a gimmick. The central character in *Gangbusters* was really the criminal, but the central character in most crime series was the detective, who was glamorous or unique. He may have been a tough guy existing on the fringes of the law, almost a criminal himself, like Sam Spade, or he might have a unique physical characteristic, like the Fat Man ("Dashiell Hammett's fascinating creature").

Mr. and Mrs. North, a sophisticated couple, fought crime. Alice Frost as Pam North and Joseph Curtin as her publisher husband Jerry attacked evil in a light-hearted manner.

But in radio crime drama's heyday, listeners followed *Ellery Queen, Philip Marlowe,* and *Martin Kane, Private Eye, Bulldog Drummond,* and *Nick Carter.* Women detectives (save for Pam North) did not exist; women served only as secretaries and assistants.

Frequently, the hero was a news man, as in *The Adventures of Christopher Wells,* or *Front Page Farrell* (which eventually became a daytime soap opera). There was *Big Town* with Steve Wilson (Edward Pawley) and his reporter Lorelei Kilbourne (Fran Carlon) who put out the *Illustrated Press. Casey Crime Photographer* always ended with Casey snapping a picture of a criminal committing a crime.

There were other detectives. *Mr. Keen, Tracer of Lost Persons* was a program of amazing popularity; in fact, it ran longer than other detective series. Mr. Keen always traced amnesiacs, runaways, or missing pets.

Mr. District Attorney began in 1939 and continued into the fifties. He seemed to have no name; he was always referred to as Mr. DA. But he (Jay Josten played the part) continued to fight crime with his sidekick Harrington (Walter Kinsella) and Miss Miller (Vicki Vola).

19 "Elementary, my dear Watson," says Sherlock Holmes (Basil Rathbone) in one of the many detective stories by Sir Arthur Conan Doyle which were presented on radio.

20 Len Doyle (left) and Jay Josten in the 1942 *Mr. District Attorney.*

21 Marion Shockley and Ted de Corsia face *Ellery Queen* (Carleton Young) in a moody studio setting. — (NBC photo)

The archetypal detective was, of course, Sherlock Holmes, and he and Dr. Watson appeared on radio over a long period of time with numerous people in the roles. The most famous, however, was Basil Rathbone with Nigel Bruce as Dr. Watson.

One of the last private eye shows to be heard on radio was *Johnny Dollar,* about an insurance investigator who specialized in insurance fraud cases, which shows you how the private eye business changed—from the general practitioner like Sam Spade (who usually dealt with murders) to those who worked for corporations.

Westerns

In addition to detective drama, Americans loved westerns. The West has always been part of the American Myth and it reached its golden age on radio with the *Lone Ranger* broadcast from the studios of WXYZ (Detroit). The masked rider of the purple plains with his faithful Indian scout Tonto lived in the imagination of every youngster who grew up in the thirties and forties. Someone once defined an intellectual as a person who could listen to *The William Tell Overture* without thinking of *The Lone Ranger* (the overture was the show's musical theme). The Masked Man stood for law and order in the Old West (and the East as well). George W. Trendle's creation, The Lone Ranger was the perfect human being, capable of doing no wrong, ingenious, sympathetic, and merciful. He was the perfect model for youngsters. When he was forced to shoot someone he always shot (usually at the last instant to prevent the gunman from killing) the gun out of his hand.

No one in the towns which he cleaned up ever bested him; the personification of justice always triumphed. The show was a ritual. It began and ended in the same way. In the final seconds, someone always said, "You know who that was? The Lone Ranger." And in the distance, you would hear, "Hi Yo Silver, Away."

Gunsmoke was another radio program which dealt with the West. It became even more popular on TV, eventually becoming one of the longest running shows of all time.

Super-Hero Thrillers

The best heroes for radio were characters like *The Shadow* and *The Green Hornet* who existed best in the imagination. Just as Superman was better imagined (the trick movie effects detracted from him) so these characters were perfect for radio and the imagination.

The Shadow is probably the more famous. Lamont Cranston was supposed to have learned the power of making himself invisible by hypnosis while in the Orient (we once thought the Orient had every answer) and he used this device to thwart his enemies. He would enter their hideouts, catch them committing a crime, and baffle them with his mysterious voice (achieved with a filter which cut out high and low frequencies just as a telephone does). Then, since they couldn't see him, they began shooting wildly, thrashing about, and otherwise trying to escape the man who knew what evil lurked in their hearts.

The Green Hornet also had a secret identity; he was Britt Reid, owner and publisher of *The Daily Sentinel.* But he would don a fancy outfit, and with Kato, his valet (Japanese until World War II, then Filipino) jump into Black Beauty, his car, and roar away after crooks. He used a gas gun (apparently painless) and left his seal (a mark with his ring) on the culprit's forehead so that the police could thank

22

23

22 "Who knows what evil lurks in the hearts of men? The Shadow knows," introduced Lamont Cranston, the first radio crimefighter to have a dual identity. Here he clouds a villain's mind so that he can't be seen.

23 William Conrad and Henry Morgan appeared on many dramatic shows.

24

him for solving the crime. Curiously the cops never did; they always suspected that he was a crook! The authorities never learned he was on their side; there is no explanation except it made the series successful.

The superheroes were alike in many ways. They wore unusual clothes and had other unusual characteristics. They acted on their own, not as part of any official law enforcement agency. Like Sherlock Holmes, they were self-trained. Since they were not members of any law enforcement agency, they would catch criminals without playing according to the rules established by courts of law. They would assault criminals without worrying about being accused of police brutality.

Too, they would eavesdrop or enter premises without a search warrant. True, they would be considered burglars if the police caught them but this only made it interesting because both the cops and the crooks were after the superhero.

The characters were almost invincible. But they worried endlessly about protecting their secret identity. In addition, some characters could be rendered powerless by various objects. Superman, for example, was weakened by kryptonite.

Shock Thrillers

Adults were drawn to mystery shows and kids enjoyed being scared by these programs too. One popular program, *I Love a Mystery,* had three major characters: Doc (a locksmith), Jack (a thinker), and Reggie (a muscleman). Together they were ready for almost anything—like dealing with witch-doctors or escaping from voodoo cults.

Inner Sanctum was the scariest of all radio shows, but it was always played tongue-in-cheek from its debut January 7, 1941. Raymond, the host, slowly opened the creaking door and began his introduction. He loved puns about horror and the macabre. Although mystery was a key part of the program, there was always the attempt to frighten the listener. The tales were about a perfect crime going wrong or someone falling into a horrible predicament through no fault of his own.

"The Coffin in Studio B" on *Lights Out* (written by Arch Oboler) had a strange little man visiting a rehearsal for *Lights Out* and attempting to sell a coffin to one of the actors. After the sale, the murder occurred while the radio show was being broadcast. Another show in the series, "The Revolt of the Worms," was about an accident in which a growth inducing chemical was spilled on the ground and was absorbed by earthworms who grew to a tremendous size, destroying the experimenter.

Suspense created many frightening programs like the one about the man who wanted to murder his wife and who asked a tough guy, Edward G. Robinson how to do it. The most memorable was the *Suspense* production of "Sorry Wrong Number" (starring Agnes Moorehead) about a woman who overhears a conversation about a murder and then is phoned by the murderer. This was later made into a film.

WE RENT RADIOS
by Day, Week or Season
Burt's Radio Parlors
DIAL 4-4757
402 ATLANTIC AVE.
OPEN EVENINGS

1931

24 "Sorry, Wrong Number" was one of the most popular radio dramas of all time. (This photo is from the film version.) — (1948 Hal Wallis Production, Inc.)

Other Drama Types

There were combination shows, too, like *The Roy Rogers Show* and *The Gene Autry Show* which were western adventure shows combined with music shows.

Further, there were the straight drama shows which usually offered adventure and thrills as well. *The Lux Radio Theatre* presented radio versions of movies using the stars who had created the movie roles. The host was Cecil B. DeMille, a director who made many biblical movie epics. This gave the show the feeling that it was a Hollywood creation, though in fact, DeMille never directed any of the shows. A studio audience witnessed only a dramatic reading.

The Theatre Guild on the Air did the same thing with Broadway plays such as *What Price Glory* and *The Milky Way* with Hollywood stars.

Most people enjoyed arriving with Mr. First Nighter at the "little theater off Times Square" just as the curtain was going up on an original half-hour play. *First Nighter* didn't come from Times Square but originated from Chicago and Hollywood. The program tried to sound authentic: an usher would direct smokers to the lobby at the end of the first act. The program ran from 1929 until 1953, setting a record as the longest running play series.

There was also the *Screen Guild Players* for radio adaptations of screen stories which usually starred actors different from those who appeared in the movie versions. *Family Theatre* was a series originated by a Catholic priest in an attempt to get family type shows on radio.

An adventure series which ran on Saturday at 12:00 noon was *Grand Central Station.* The activities had to do with New York City and people arriving or leaving it.

History was popularized with *Cavalcade of America.* Lionel Barrymore starred in the *Mayor of the Town.* Although not really an adventure program, *Dr. Christian* offered great interest. Helen Hayes appeared on radio adaptations of her stage hits in *The Helen Hayes Theatre.*

Plays written especially for radio were to be heard on *The Columbia Workshop* which featured famous writers, *The Free Company Presents,* and *This Is Corwin,* as well as *Words Without Music,* a program which was just that.

"War of the Worlds"

No question that the most famous adventure program of all time was the one broadcast on Sunday night, October 31, 1938: *The Mercury Theatre of the World* (produced and directed by Orson Welles). This was an experimental show titled "War of the Worlds." It dealt with the invasion of the earth by Martians. What made the program unique was that it used a realistic news format with people speaking from supposed remote locations, breaking in on one another, making mistakes, and interrupting regular programs just as radio listeners had been conditioned to expect in real emergencies.

The results were unexpected. People who forgot to check other stations panicked. Families bundled into cars and tried to drive out of the cities clogging the roads. People ran out of their homes and stood in the streets looking up at the sky. The cast in the CBS

25

25 Sidney Poitier appeared on *Dimension X* and *X-1,* science fiction shows. Sci-fi seemed perfect for radio because its content was better imagined than seen. Unfortunately, science fiction broadcasts came on the air in the waning days of radio drama. Even so, they found a loyal audience. — (National General Pictures)

1941

1943

studios was not aware of the havoc they had wreaked until after the show. The nation was so disturbed that the young actor-producer-director Orson Welles was called before a congressional committee which was investigating the incident.

It all proved that radio was a medium of great power especially in its ability to convince people of something they could not prove by use of any of their other senses. Further, it indicated that the public had tremendous faith in radio as a truthful medium. It accepted unquestioningly what went out over the air as fact.

Finis: Radio Drama

With the advent of TV in the late forties and early fifties, radio drama died; it couldn't compete with visually dynamic dramas like the ones seen on *Playhouse 90, The Schlitz Playhouse of Stars,* and *The Kraft Theatre.*

Gangbusters

"The Eddie Doll Case"

This was a dramatized true crime adventure program. It was created by Phillips H. Lord.

Narrator—Col. H. Norman Schwarzkopf
Announcer—Frank Gallop
Director—Jan Hanna
Producer—Phillips H. Lord
Writer—Bruce Disque, Jr.

The cast included (at times) Art Carney, Richard Widmark, Frank Lovejoy and others.

COLONEL.—Dr. Simon, I understand that tonight's case concerns Eddie Doll, alias Eddie Larue, alias Burlington Eddie, alias Edward Foley.

SIMON.—Yes, Colonel Schwarzkopf. The case starts on September 16, 1930. Late at night, in the gang's hide-out at Lincoln, Nebraska, a barren room in the back part of a dilapidated apartment house. The shades were drawn, the windows sealed. The room was stuffy . . . blue with smoke . . . a tenseness was in the air. The gang was waiting nervously.

SOUND.—*Sneak in footsteps walking back and forth.*

ROGERS.—Sit down, Buck and take a load off yer feet!

TIM.—Yeah. You gimme the willies walking around!

ROGERS.—Forget it. Wait until Eddie Doll gets here. He's got all the low-down.

TIM.—I'm glad Doll has joined up with the gang. He's got a cool head . . . He's slick . . . He ain't one of these guys that goes off half cocked.

BUCK.—This is going ter be the smoothest bank job ever pulled in this country!

SOUND.—*Three knocks . . . two knocks.*

BUCK.—There's Doll now . . .
(*Half fade*)

TIM.—Make sure before you unbolt that door.

SOUND.—*Footsteps under Tim's line.*

ROGERS.—Who is it?

DOLL.—A guy.

SOUND.—*Slip bolt . . . door opens and closes . . . footsteps.*

DOLL.—Hi. . . .

SOUND.—*Subdued gang greets Doll.*

TIM.—Hi, Doll.

DOLL.—*(Moving chair)* Everybody here?

BUCK.—Yeah . . . When we going ter crack the job?

DOLL.—We're going ter crack it in the morning, boys.

SOUND.—*Slight reaction.*

ROGERS.—What's the dope?

DOLL.—This is going ter be the most perfect bank cracking ever pulled in this country . . . and the biggest. One million dollars.

BUCK.—Everything's set.

DOLL.—We're going to rehearse this thing inch by inch right now. I've got every emergency covered. . . .

BUCK.—We been working on it 3 months.

DOLL.—Buck, you're responsible for the getaway. If anything slips we'll all get lead poisoning. . . . Give the boys the setup.

BUCK—Here's the map of our getaway. *(Sound of paper)* The car will do 70. You guys jumps in. . . . We heads north, take the turn into Elm Street. . . . They're working on the road, so only one car can pass. I'm giving a taxicab driver 100 bucks. After we pass, he starts to drive through . . . stalls his car . . . so anybody chasing us will be stuck.

ROGERS.—Does he know what we're up to?

BUCK.—Of course not. . . . I told him it was a wedding party trying to get away. . . . We keeps going . . . on the state highway over the railroad track. We turns off the highway, here, and heads for the hangout. I've drove over that route three times a day for the past two weeks. I could drive it blindfolded.

TIM.—How about license plates?

BUCK.—I got that fixed, too. . . . While we're driving we can drop off the license plates . . . and swing new ones on. O.K., Doll?

DOLL.—All right. Now . . . fer weeks I've had all you guys going in the bank . . . having money changed. . . . I hired a vault terday . . . got a good look through the cellar. *(Rustle of paper)* Here's a picture of the inside of the bank. There'll be four guns in the bank . . . one

1941

in that drawer there. . . . and the two guards are always standing right here in front of this cage. Frank . . .

FRANK.—Yeah!

DOLL.—You stands to the right of *this* guard, and Tim stands to the right of *this* one. . . .

TIM.—O.K.

DOLL.—At the signal . . . you crack them guards . . . snatch their guns. I'll get the two guns from the drawers. Ten seconds later, Seeney comes in the bank with a machine gun. He covers the customers. *(Ad lib agreement)* We touch nothing in the bank but money. . . . *We leave no fingerprints.* Remember that!

BUCK.—Suppose the two guards puts up a fight.

DOLL.—Frank and Tim bumps 'em off. Now . . . let's study the layout. We'll spend the rest of the night *memorizing every detail.* *(Fade in)*

SOUND.—*Slight bank commotion . . . adding light background.* *(Fade in)*

MAN.—Good morning, Mr. Smith. . . . I'd like to cash this check.

TELLER.—Certainly, Mr. Brown.

DOLL.—Stick 'em up. . . . This is a holdup! *(Quiet)*

DOLL.—Number three . . . keep 'em covered with that machine gun. If anybody makes a move, mow the whole bunch down.

ROGERS.—Right.

DOLL.—Number one . . .

BUCK.—*(Half off)* Yes, sir . . .

DOLL.—Scoop all the loose cash into those laundry bags. Hey, you . . . you bank guy! Come here. Come with me and swing back the door of that vault.

BANK GUY.—Yes, sir,

SOUND.—*Footsteps.*

DOLL.—Swing it back.

SOUND.—*Several bolts.*

DOLL.—One false move and you'll get lead poisoning . . . Scoop all those bank notes into that bag.

SOUND.—*Much change being poured into bags . . . many packages of bills being tossed in . . . continues under* *(Fade in)*

ROGERS.—There's a lot of loose money in these drawers, boss.

DOLL.—Take yer time, Pal. . . . This ain't no peanut robbery. Keep cool. . . . Use yer heads. This is going to be a million dollar haul. . . . Lug the full bags of money as far as the front door.

SOUND.—*Change going into bags out.*
(Fade in)

BUCK.—There's a crowd collecting in front of the bank.

DOLL.—Let them collect. . . . We're collecting in here.

ROGERS.—*(Coming on)* We got everything, boss.

DOLL.—You guys carry those bags. *(Projected)* Don't one of you people take one step to follow us . . . or *we'll* shoot holes in you. Come on . . . *(footsteps)*

DOLL.—Number three . . . you stand at the door with the machine gun. As soon as we're all in the car, we'll give you the signal.

ROGERS.—Check.

DOLL.—Come on. *(Footsteps . . . crowd gets louder . . . sound of motor idling)* Throw the bags of money in the back. *(Several thuds)* Sound the signal . . . Everybody get in. *(Horn two long blasts)* Here come the rest of the gang. . . . *(Car door slams)* Step on it, Buck.

SOUND.—*Roar of motor up strong and fade out.*

COLONEL.—As I recall, Dr. Simon, that million dollar robbery was the biggest bank robbery ever staged in this country. Please tell our Palmolive Shave Cream listeners what happened next.

SIMON.—At that time, Colonel, bank robbery was not under the jurisdiction of the Federal Bureau of Investigation, but the Lincoln, Nebraska, authorities asked the F.B.I. to furnish them with what information it could. On September 19, 1930, Inspector Haynes of the Federal Bureau of Investigation was in his private office in Washington, and I want you to see how a large organization operates in gathering complete information about a criminal. . . .

SOUND.—*Door opens and closes.*
(Fade in)

DENNISON.—You sent for me, Inspector Haynes.

HAYNES.—Dennison. This bank robbery in Lincoln, Nebraska, is the cleanest bank job ever done in this country.

DENNISON.—There isn't a clue. . . .

HAYNES.—Not one . . . *but* . . . we've *got* to *find* one. Now I've got here a complete report of how the robbery was executed. I want to check the *modus operandi* of this gang.

DENNISON.—Every musician has a definite individual musical touch. . . . Every painter has his own style. . . . Every criminal has his own individual approach to a crime.

HAYNES.—Yes, Dennison. We know there are some dozen gangs of bank robbers in the Middle West. . . . We know some recent gangs have been broken up. . . .

DENNISON.—You mean this robbery may have been committed by a leader who had his schooling from some gang that's already been caught?

1929

1941

1927

1946

1942

1942

HAYNES.—That's it. And if we can find a Middle Western gang, which operates similar to the procedure used in this robbery, it'll be a nail to hang our hat on.

SOUND.—*Buzzer . . . click.*

HAYNES.—Yes?

FILTER 1.—Mr. Frank is here, Inspector.

HAYNES.—Send him in at once.

FILTER 1.—Yes, sir.

SOUND.—*Click.*

HAYNES.—*(To Dennison)* I've asked several of the men to get reports on some of these Midwestern gangs, Dennison. . . .

SOUND.—*Door opens . . . closes.*

FRANK.—Frank reporting, Inspector.

HAYNES.—Get a report on that Salta gang?

FRANK.—Yes, sir. . . . They pulled four bank robberies. In each case, they shot the guards, and in each case they were careless about fingerprints . . . and they didn't bother to take along loose silver.

SOUND.—*Buzzer . . . click.*

HAYNES.—Yes?

FILTER 1.—O'Brien is waiting, Inspector.

HAYNES.—Tell him to come in.

FILTER 1.—Yes, sir. . . . That's all, Frank.

FRANK.—Yes, sir.

SOUND.—*Door opens . . . closes.*

HAYNES.—*(To Dennison)* Dennison . . . the Salta gang had no part in this bank robbery. This gang we're looking for *scooped up the loose silver. . . .* They left no *fingerprints. . . .*

SOUND.—*Door opens and closes.*

DENNISON.—And the leader of the gang we're chasing had *brains.*

O'BRIEN.—*(Coming in)* O'Brien reporting, sir.

HAYNES.—What about that Hoosier gang, O'Brien?

O'BRIEN.—There are six robberies laid to them, Inspector. In all six cases they entered the bank, bound their prisoners, took all money . . . *including loose silver.* Only in one case was there *shooting of a guard.* The gang wore gloves in all cases, and only in one instance did they desert their car.

HAYNES.—Thanks, O'Brien. . . . Is Smith waiting outside?

O'BRIEN.—Yes, sir.

HAYNES.—Ask him to come in.

O'BRIEN.—*(Fading)* Yes, sir.

SOUND.—*Door opens and closes.*

HAYNES.—Dennison . . . this gang sounds more like the one we're after. . . . their not shooting guards agrees. . . . Not deserting their car agrees. . . . Taking all loose silver agrees. . . .

DENNISON.—But the tying up of all those in the banks. . . .

SOUND.—*Door opens and closes.*

HAYNES.—That's where the Hoosier method differs from the gang that pulled this job. Hello, Smith. . . .

SMITH.—*(Comes on)* I got reports on the Five-finger Mob and on the Yates gang.

HAYNES.—Let's have them.

SMITH.—Three bank robberies during the past year have been laid to the Five-finger Mob. In every case they've been scared off.

DENNISON.—Isn't that the gang that always leaves a girl in the car out front as a blind?

SMITH.—Yes, sir.

HAYNES.—Um . . . they're just a rattlebrained mob. But there was a *super thinking mind* in back of this Lincoln, Nebraska, job. What about the Yates gang?

SMITH.—Broken up about two years ago. Three of them caught . . . one shot. . . . *Yates and two others escaped.* Nothing has been laid to this gang during the past 2 years.

HAYNES.—What was their procedure?

SMITH.—*(Rattle of papers)* I have it here. They entered the bank . . . made a thorough sweep of money . . . held employees at machine gun point. . . . In four cases they got reserve money from vaults, had a car waiting to make escape. In no case did they ever desert the car.

HAYNES.—Thanks, Smith. That's all.

SMITH.—Yes, sir.

SOUND.—*Steps . . . door opens and closes*

DENNISON.—Inspector, that's the same *modus operandi* used in the Lincoln, Nebraska, robbery.

HAYNES.—Yes, Dennison. *(Rustle of paper)* This report says that back in February our St. Louis field office was notified that a sheriff in Macomb, Illinois, picked up an Edward Doll for stealing a car. Doll was arrested, placed under $3,500 bail . . . skipped bail. A car thief doesn't usually have $3,500 to put up as bail . . . or to throw away by not appearing.

SOUND.—*Dictograph.*

HAYNES.—Fingerprint Department?

FILTER 1.—Yes, sir.

HAYNES.—Look up the record of Edward Doll . . . See who he's been connected with in the past.

FILTER 1.—Yes, Sir.

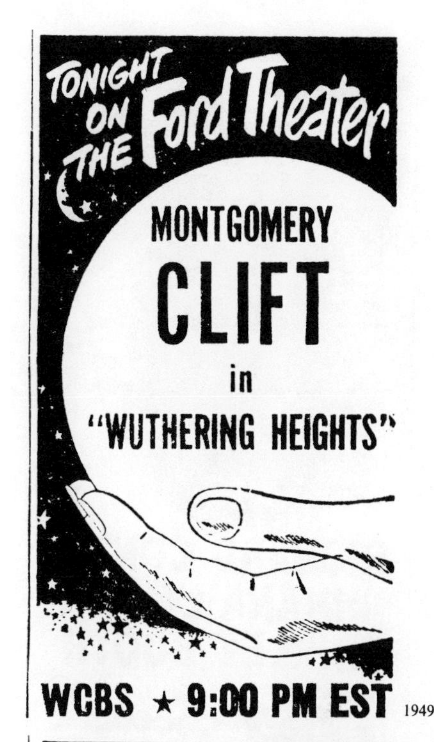

1949

1943

1949

SOUND.—*Dictograph click.*

DENNISON.—Are you figuring that Doll may have joined up with the *Yates gang?*

HAYNES.—Let's think now. The Yates gang always makes a thorough cleaning of the bank. That tallies. They don't tie the customers and employees. That tallies.

DENNISON.—And their general plan of procedure was similar to the procedure just used in this bank robbery.

HAYNES.—All right. . . . Yates and two of his pals are still at large. . . . Edward Doll skipped his bail 4 months ago. . . . It's taken several months to plan this bank robbery. . . . Why do all of these facts fit so perfectly?

SOUND.—*Dictograph buzz . . . click of receiver.*

HAYNES.—Inspector Haynes speaking.

FILTER 1.—Report on Edward Doll . . . arrested several times on minor offences . . . known to be exceptionally clever. . . . It is possible that he is one of the leaders reorganizing a Western bank robbery gang. That's all.

HAYNES.—O.K. . . .

SOUND.—*Dictograph click.*

DENNISON.—So Doll *does* know Yates?

HAYNES.—Of course some of this is hypothetical, but Doll may have learned his bank robbing from Yates and now is even more clever. See that a picture of Doll, his history, and all of these facts are sent to the Nebraska authorities. It may be a good lead.

SIMON.—A number of months went by, Colonel . . . Doll completely disappeared. Then suddenly . . .

FILTER 2.—Kidnaping . . . South Bend, Indiana. Kidnaping corresponds to description of Edward Doll, recently distributed by the Federal Bureau of Investigation.

FILTER 1.—Local bank in Tupelo, Mississippi, just robbed by Machine-gun Kelly. Description of his companion fits Edward Doll. Wanted . . . all information on Machine-gun Kelly and Edward Doll.

FILTER 3.—Attention . . . attention. . . . United States mail robbery at Effingham, Illinois. . . . Believe leader of gang was Edward Doll.

CHIEF.—Well, Colonel, orders came from headquarters to the G men to redouble their efforts to get Edward Doll. Inspector Haynes again called Dennison into his office in Washington.

HAYNES.—Dennison, all we know up to now is Edward Doll is somewhere in this country . . . and we've got to find him. I've just received some additional information.

DENNISON.—Something I haven't heard?

HAYNES.—Yes. Reports from all over the country. Doll's weakness is pretty girls. Has been known to visit Alice Kahn of New Orleans, Mildred Barling of San Francisco, Lucy Weber of Denver, Joan English of St. Louis. . . . There's a long list of them here, sir.

SOUND.—*Dictograph buzz . . . click.*

HAYNES.—Inspector Haynes.

FILTER 1.—Report on Doll, sir. Used to pal around with Kathryn and Machine-gun Kelly. We have his fingerprints . . . specimens of his handwriting, and a good picture of him.

HAYNES.—Send it all in immediately.

SOUND.—*Click of dictograph.*

HAYNES.—Well, Dennison, we're not very far ahead.

DENNISON.—Wait, Inspector, here's information I've dug up. Doll is fond of motion pictures . . . especially gangster pictures. And get this. . . . He has *tattoo marks on both his right and left forearms.* A heart and an anchor and figure of a girl on his right forearm . . . a cowgirl and pierced heart on his left forearm. And he's quoted as having said that eventually he's going to retire to a chicken farm.

HAYNES.—Well, now we're getting somewhere! *(Dictograph buzz . . . click)* Inspector Haynes.

SMITH.—*(Filter)* Men at one of our Western field offices have just been to the jail and talked with Machine-gun Kelly. They didn't let Kelly know they wanted information on Doll . . . but they asked a lot of questions about other things, and Kelly intimated that about 9 months ago Doll married a girl by the name of *Janet Galaton* in New York.

HAYNES.—Thanks, Smith.

SOUND.—*Click.*

DENNISON.—Doll's *already* married, Inspector.

HAYNES.—That wouldn't stop *him* from marrying again. Um . . . that's the best tip we've had yet.

DENNISON.—It's going to be a big job to examine all the marriage licenses in New York. They probably got married under fictitious names, too.

HAYNES.—*(Thinking)* Let's see. . . . Doll meets a girl . . . wants to marry her. Is he going to let her know who he *really is?* If she were the type of girl that he could take in as one of the gang . . . he wouldn't care if she knew his identity. Right, Dennison?

DENNISON.—Yes.

HAYNES.—*But* . . . if Doll *doesn't want her to suspect anything* then he can't ask *her* to sign a fictitious name to the marriage license. He'd change *his* name . . . but that marriage license is going to contain *her real* name. We've got to find a license made out to *Janet Galaton.*

COLONEL.—Dr. Simon, I know the most interesting part of the case will be the police search for Janet Galaton, but before you tell us about that, Frank Gallop has a few words for our listeners. *(Commercial)*

COLONEL.—Dr. Simon, you were telling us that Inspector Haynes and Dennison went to New York to check the marriage license records.

RADIO FACTS!

Crime Show Introductions!

Friend to those who need a friend; enemy to those who make him an enemy.
Boston Blackie

Out of the night, out of the fog, and into his American adventures comes Bulldog Drummond.
Bulldog Drummond

Click! Got it! Look for it in the *Morning Express!*
Casey, Crime Photographer

He hunts the biggest of all game — public enemies who try to destroy our America!
The Green Hornet

And it shall be my duty as District Attorney, not only to prosecute to the limit of the law all persons accused of crime perpetrated within this county, but to defend with equal vigor the rights and privileges of all its citizens.
Mr. District Attorney

SIMON.—Yes, Colonel Schwarzkopf, for over a month, Inspector Haynes and Dennison worked in the marriage bureau of New York, carefully going over every marriage license. It was a tedious job. *(Fade)* Then, late one afternoon . . .

SOUND.—*Typewriters and office hum in background.*

DENNISON.—*(Tense)* What is it, Inspector Haynes?

HAYNES.—This is worth our weeks of tiresome checking, Dennison. A marriage license made out to Miss Janet Gabrielle Galaton and Mr. Leonard E. Foley.

DENNISON.—Janet and Galaton are fairly common names.

HAYNES.—Married to Leonard E. Foley. . . . Give me that sample of Doll's handwriting.

DENNISON.—Sure, here it is.

SOUND.—*Rustle of paper.*

HAYNES.—Let's see. . . . Leonard E. Foley. . . . Um . . . Edward Doll. Those two E's are written exactly the same way.

DENNISON.—Yes . . . the ends on those two D's are the same too . . . and notice the R's?

HAYNES.—Dennison, it's our first clue!

DENNISON.—Foley gives his address there as the Nemo Hotel, Dallas, Texas. . . . That's probably fictitious.

HAYNES.—But Janet Galaton gives *her* address as Danville, Vermont. That's probably correct. Come on, Dennison. We're going to Danville, Vermont.

SIMON.—Several hours later, Colonel, Inspector Haynes, and special agent Dennison, posing as traveling salesmen, arrived at Danville, Vermont. They rented a car and drove up to the local post office.

SOUND.—*Car stopping.*

DENNISON.—But why come to the post office, Inspector? Let's go up to her home.

SOUND.—*Turn motor off.*

HAYNES.—This is safer.

DENNISON.—Why?

HAYNES.—Janet Galaton doesn't know she's married to a man like Doll, does she?

DENNISON.—No.

HAYNES.—If she doesn't know, *her parents* don't know.

DENNISON.—Um . . .

HAYNES.—It's natural for parents to communicate with their daughter. If they don't realize there is any need for secrecy the *most natural* communications would be through the mails.

DENNISON.—I get you.

HAYNES.—The safest way to close down on a criminal is not to let anyone in the world know you're looking for him. Come on . . . in the post office.

SOUND.—*Door of car . . . footsteps on cement . . . change footsteps to wood.*

HAYNES.—*(Fade in)* How do you do.

POSTMASTER.—How do you do, sir.

HAYNES.—I'd like five 3-cent stamps.

POSTMASTER.—Yes, sir.

SOUND.—*Money on counter.*

HAYNES.—Say . . . by the way. . . . Would you direct me to the home of Janet Galaton?

POSTMASTER.—Yes, sir. You turn right . . . a mile up the road . . . a green house on the right-hand side.

SOUND.—*Sealing letters.*

HAYNES.—Thanks. Thought I'd drop in and surprise her.

POSTMASTER.—She's not at home now, you know.

HAYNES.—*(Disappointed)* Doesn't she live here any more?

POSTMASTER.—*(Laugh)* Hasn't for a year. Married now . . . married some out-of-town fellow.

HAYNES.—I'm awfully disappointed. Say . . . I guess I'll drop her a postal card from here, though. You don't happen to know her address offhand, do you?

POSTMASTER.—Yes. . . . Her folks sent her a package yesterday and insured it. . . . The slip . . . here it is. The package was sent to Mrs. Janet G. Foley, Box 270A, Route No. 2, St. Petersburg, Florida.

HAYNES.—Thanks. I'll drop her a card. *(Slight pause)*

COLONEL.—That was a clever piece of work, Dr. Simon. Please tell our Palmolive Shave Cream audience how Inspector Haynes followed it up.

SIMON.—Well, Colonel, 4 days later, Inspector Haynes and special agent Dennison arrived in St. Petersburg and talked with the local post office officals.
(Fade in)

HAYNES.—But I tell you there *must* be a Leonard E. Foley listed somewhere here in St. Petersburg.

POST OFFICER.—*(Florida accent)* No Leonard E. Foley in the city directory.

HAYNES.—Is the postman who delivers over Route No. 2 around?

POST OFFICER.—He may be in the other room. *(Fade)* I'll see.

HAYNES.—*(Lower voice)* What do you think, Dennison?

DENNISON.—Funny there isn't a Leonard Foley in the directory.
(Fade in)

1949

1949

Post Officer.—Mr. Jenkins was just going out delivering mail. He has Rural Route No. 2. These two gentlemen are Federal officers, so answer anything they ask you.

Postman 2.—*(Florida accent)* Yes, sir.

Haynes.—We have the address of a Leonard E. Foley, Box 270A, Route No. 2. Know anything about him?

Postman 2.—Why . . . about 2 weeks ago he wrote out an order that all mail addressed to Foley should be delivered . . . Wait a minute. . . . I've got that order here in my book.

Haynes.—Good. . . . Did he write the instructions himself?

Postman 2.—Yes, sir.

Haynes.—Dennison . . . let me have that sample of Doll's handwriting.

Dennison.—Just a minute.

Sound.—*Shuffle of cards.*

Postman 2.—Here's the note he wrote out. Says to deliver any mail addressed to Janet or Leonard Foley to 5190 38th Avenue North.

Haynes.—Let me see the paper. . . . Um . . . Dennison . . . notice this E. . . . See this R. . . .

Dennison.—That's his handwriting!

Haynes.—He and his wife have a house out there?

Postman 2.—Yes, sir. Farm about 35 acres. They raise *chickens.*

Haynes.—When's the last time you saw him?

Postman 2.—He come out to the postbox about 2 days ago.

Haynes.—Thank you, gentlemen, very much. Come on, Dennison, we'll go out and see this Mr. Foley!

Sound.—*Slight pause . . . motor fading in.*

Haynes.—That looks like the house, Dennison, ahead on the right.

Dennison.—Think there'll be shooting, Inspector?

Haynes.—There will be if he can get his hands on his gun . . . but first we've got to make *sure* he's the *right man.*

Dennison.—Remember the tattoos. He should have a heart and anchor and a girl tattooed on his right forearm. . . .

Haynes.—But on his forearm . . . if we try to force him to roll up his sleeves there may be shooting.

Dennison.—That would be a sure way to identify him, though.

Sound.—*Car slows up.*

Dennison.—There's a man around back of the house.

Sound.—*Car stops.*

Haynes.—*(Low)* Change your gun into your outside pocket.

Dennison.—*(Low)* Right.

Sound.—*Car door opens . . . feet in straw walking.*

HAYNES.—*(Calling)* Hello, there. Mind if we come out back and see you?

DOLL.—*(Distance)* Come on.

SOUND.—*More walking.*
(Fade in)

HAYNES.—We're interested in buying some chickens. . . .

DOLL.—*(A little surprised)* You ain't *farmers*. . . .

HAYNES.—No . . . we've just moved to St. Petersburg. Thought we might arrange to get fresh chickens from you.

DOLL.—I ain't selling any. Wait a minute till I close that gate. *(Fade)* All the chickens will be out.

DENNISON.—*(Whisper)* Think it's Doll? He's about the right size . . . heavier, though.

HAYNES.—*(Whisper)* We got to get a look at his *forearm.*
(Fade in)

DOLL.—What did you two fellers stop for, anyhow?

HAYNES.—I told you. . . . We'd like to have you kill us a fresh chicken every Sunday. You've got a nice place here. . . . This big hogshead makes a good watering trough.

DOLL.—Yeah. . . . It's always full of water, too. . . . This hose runs from that spring over there and keeps the hogshead full.

HAYNES.—Look here, Bill. . . . Lean over and look in it. . . . Isn't that water clean?

DENNISON.—Mighty clear.

SOUND.—*Drop watch in water.*

HAYNES.—Oh! I dropped my watch into the water! You've got your coat off, sir . . . would you roll up your sleeves and get it out before the water gets into the works?

DOLL.—All right. . . . Wait a minute. . . . *(Bending over and grunt . . . swish of water)* There. . . .

HAYNES.—Thanks. . . . It was awfully careless of me.

DENNISON.—Got a tattoo mark on your arm, haven't you? An anchor and a girl. . . . You must have been a sailor.

DOLL.—No. . . . I did it for the fun of it.

DENNISON.—This is interesting. . . . Tattoo always fascinated me. . . . Let me see it. . . .

DOLL.—Sure.

HAYNES.—That's beautiful work.

DENNISON.—Best I've ever seen!

DOLL.—You fellas think *this* is good? Wait til you see my *other* arm!

HAYNES.—*(Leading him on)* Oh, you've got *another* tattoo?

DOLL.—*(Proudly)* Look at *this!*

1948

CLAUDE RAINS
JUNE DUPREZ
GEORGE COULOURIS
DEAN JAGGER
IN
"Valley Forge"
You'll find inspiration and entertainment in the thrilling story of how George Washington found faith and courage in the darkest hour of the Revolution.
Theatre Guild on the Air
WJZ - 9:30 P. M.
presented by
UNITED STATES STEEL

1948

DENNISON.—A cow girl and a pierced heart.

HAYNES.—Hold your arms out together so I can compare the two designs.

DOLL.—Sure. What do ya think of 'em?

SOUND.—*Sudden click of handcuffs.*

DOLL.—*(Astonished)* Hey. . . . What's the idea . . .? Take these handcuffs off me!

HAYNES.—Edward Doll, you're wanted for the million dollar bank robbery in Lincoln, Nebraska; the South Bend kidnaping and too many other crimes to mention.

DOLL.—So you *know* me? How'd you find me? . . . I didn't make *one* false move.

HAYNES.—That's one of the things that *helped* us find you, Doll.

SIMON.—And that, Colonel, was the end of Eddie Doll, master mind of the biggest bank robbery ever committed in this country.

COLONEL.—What happened to Doll, Dr. Simon?

SIMON.—He was sentenced to a long term in a Federal penitentiary.

Have you tried Wheaties?
They're whole wheat with all of the bran.
Have you tried Wheaties?
For wheat is the best food of man.
So just buy Wheaties.
The best breakfast food in the land!

COLONEL —Thank you, Dr. Carleton Simon, for telling us this gripping case. You and I know that no matter how cunning a criminal may be, no matter how cleverly he may cover his tracks, sooner or later he is bound to be uncovered and suffer the full penalty of the law. Tonight's case has brought out vividly our oft-repeated statement . . . THE CRIMINAL CANNOT WIN.

1929

TUNE IN TOMORROW

Chapter 4

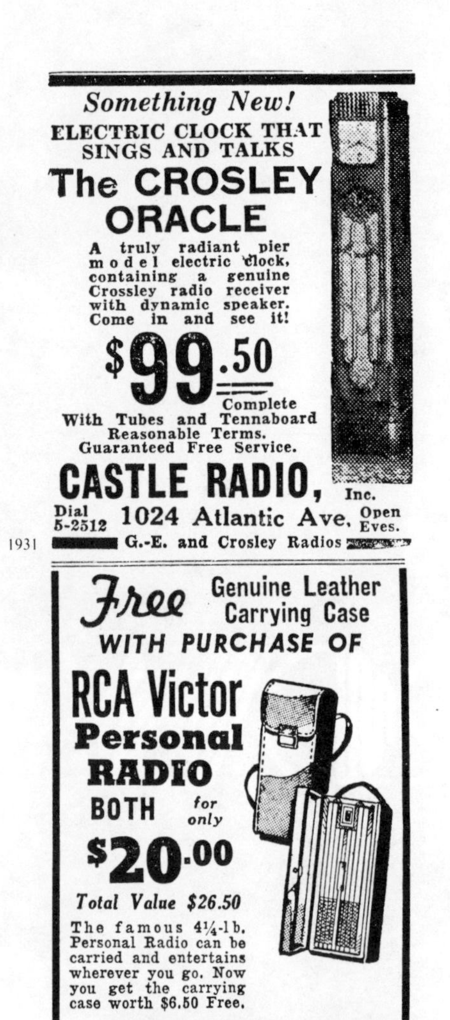

1931

1941

Announcer: Presenting Hop Harrigan. America's Ace of the Airways.
(Sound of airplane motor)
Hop Harrigan: CX4 calling Control Tower. CX4 calling Control Tower. Standing by.
Control Tower: Control Tower back to CX4. Wind southeast—ceiling 1200. All clear.
Hop Harrigan: O.K. This is Hop Harrigan coming in.

During the 1940s, these words sounded in millions of homes each weekday night at five o'clock. As mother prepared supper in the kitchen, kids sat next to the radio in the living room and listened to serials like *Hop Harrigan.*

These fifteen-minute programs began at five o'clock and followed one another until six. Although they were designed for boys, girls listened too. There was at least one program designed for girls—*Little Orphan Annie.* In addition, a woman (The Dragon Lady) was the villainess of *Terry and the Pirates.*

During wartime (1941-1946), fliers like Hop Harrigan were popular. So too were Don Winslow of the Navy and Terry and the Pirates. All these shows were about servicemen fighting the enemies of the United States.

Early Heroes

The first radio hero for kids was Buck Rogers, a spaceman, who went on the air in 1931. *The Adventures of Frank Merriwell* (a schoolboy sports hero) started in 1933. So did *Jack Armstrong.* Both *Superman* and *Captain Midnight* began in 1938.

Comics and Radio

Many characters were adapted for radio from the comic strips: Buck Rogers, Little Orphan Annie, Terry and the Pirates, Dick Tracy, Jungle Jim, Superman, Don Winslow, Smilin' Jack, Red Ryder, Mark Trail. The reasons for their selection are not clear. Apparently, it was only luck that some comic-strip characters were adapted for radio.

A few characters were created just for radio. Hop Harrigan is an example (although afterwards, he appeared in comic books). Another is *Tennessee Jed* which was a series which took place in the Old South in the years after the Civil War ended.

Some were "spinoffs" from other popular radio programs: the teenage detective, Chick Carter, was a character on a show called *Nick Carter* (about his father). He proved so popular that he got his own show. Other characters were adapted from books such as those about Frank Merriwell.

Types of Listeners

Shows in the early days had fewer listeners among the younger set because during the Great Depression (around 1933—when most men were out of work), few families could afford to buy radio sets. Too, they often felt they shouldn't permit their children to listen to radio programs when the children should have been doing their homework.

When business boomed, men went back to work and cheaper radios came on the market, and there was an increased interest in radio shows. Thus, serials like *Frank Merriwell* and *The Lone Ranger* reached the air. Jack Armstrong really caught the interest of children.

Jack Armstrong

Who was Jack Armstrong? Why, the All-American Boy! He and his sister Betty got into all sorts of adventures and mishaps fighting villains like the Black Vulture, the Silencer, and other enemies of the United States. Jack was the ideal of what many boys wished to be—dauntless, self-reliant, courageous. And he didn't have to be home in time for supper.

In one episode, a vulture is screaming while an announcer explains it as a symbol of dishonesty. We hear a train's whistle and find Jack, Betty, Billy, and Uncle Jim are on board. All this occurs with a few seconds of dialogue and sound effects which set the scene in our imagination.

They arrive in Chicago on the lookout for the Black Vulture who is a black market operator selling illegal goods.

Uncle Jim: Watch that briefcase. Once they grab it, it will probably pass from one hand to another like a football. Keep your eye on it.

We then find that Uncle Jim wants the Vulture's gang to steal his briefcase so that he can find the Black Vulture. Uncle Jim and the kids get off the train and start a fake fight. The Vulture's gang breaks in and gets the briefcase. Jack and Billy follow as the gang escapes in a taxi. Jack, however, has followed Uncle Jim's instructions and he keeps his eye on the briefcase. It is hidden under papers at a newsstand. He realizes the news dealer is in with the gang!

Jack: We have to carry the ball from here on, Billy, and it's gonna be pretty dangerous. Believe you me.

The program ends with the announcer telling listeners to tune in tomorrow to hear more about this dangerous situation.

Superman

After commercials, kids would hear the next program, which was *Superman.* This had the most famous introduction of them all:

Voices: Faster than a speeding bullet.
More powerful than a locomotive.
Able to leap tall buildings in a single bound.
Look up in the sky.
It's a bird.
It's a plane.
It's Superman.

After a commercial for Pep (a breakfast cereal), the announcer gives a brief account of what had happened and the action starts.

In one episode, Clark Kent and Jimmy are in Illyria where Herkemene (a former ruler) had been sentenced to hanging.

Kent starts towards the scaffold. A guard stops him. As always, Kent has to change into Superman and Jimmy must not see him make this change. Jimmy runs away and the guard splits his sword on Clark Kent's chest, not knowing that he is Superman. Kent changes clothes and as Superman, his voice changes too. It becomes lower halfway through when he says, "This is a job for . . . Superman." People shout as he takes the noose from Herkemene's neck.

Superman: Now don't be frightened now, Herkemene, I'm going to fly away with you.
Herkemene: I won't.
Whoosh!
Superman: Up and Away!

26

26 Ruth Yorke and James Meighan *(Marie, the Little French Princess)*

27

In radio, it was necessary to make sure the audience understood what was going on, since they couldn't see the action. To make clear what was happening, the announcer or an actor would explain what he was doing as he did it. Thus, Superman would shout "Up and away" to indicate that he was flying.

Other Programs

There was lots of competition. In the forties, each network broadcast serials. Instead of listening to *Superman* on MBS (Mutual Broadcasting System) you could hear *Terry and the Pirates* on ABC (American Broadcasting System) at the same hour. Instead of *Jack Armstrong* on ABC, you could listen to *Captain Midnight* on MBS.

Every popular serial had youngsters as the main characters or as assistants to adult heroes. For example, Dick Tracy had Junior, Superman had Jimmy Olsen, and Captain Midnight had Joyce Ryan and Chuck Ramsey.

Scriptwriters used girls to attract girl listeners. In addition, kids of varying ages were selected so that children of all ages would want to listen.

Most of the heroes were adults. They usually had glamorous jobs. Pilots and cowboys were the most popular.

Captain Midnight

Captain Midnight began with a clock sounding 12 o'clock midnight. Next came the sound of an airplane diving and the voice of the announcer shouting "Captain Midnight."

The hero of this show was a daredevil pilot who had successfully returned from a mission at the stroke of midnight. (Until that time he had been named Captain Albright.) He had young assistants named Chuck Ramsey and Joyce Ryan and a mechanic named Ichabod Mudd. His most famous enemy was Ivan Shark who headed a crime syndicate with his daughter Fang.

After each program, listeners enjoyed decoding secret messages with their Captain Midnight decoders. The code wheel used combinations like this:

J K L M N O P Q R S T U V W X Y Z A B C D
1 2 3 4 5 6 7 8 9 10 11 12 13 14 15 16 17 18 19 20 21

E F G H I
22 23 24 25 26

The announcer read a message to be decoded like this. (Go ahead. Try to decode it.)

20-18-5 16-6-12 9-22-18-21 11-25-26-10

Coded messages made kids feel as if they belonged to a show and they kept tuning in, which made the sponsor very happy.

Tom Mix

Tom Mix Ralston Straightshooters was a tremendously popular program in the forties. Tom Mix used give-away offers to attract attention to his show. Mix, who in radioland lived on the T-M Bar Ranch and rode a horse named Tony, was a movie star who had been a real cowboy and had lived a very adventurous life, having been blown-up, shot twelve times (once in a real stagecoach robbery) and while working as a stunt man, injured forty-seven times. Although he was to star on the radio show, he unfortunately died in a car accident before the radio show began. Curley Bradley played the part.

27 Tom Mix, as a radio cowboy, is well remembered by many adults who sent away for his Atomic Bomb Ring, his Mystery Ring, Decoder Badge and other give-aways.

With characters like Sheriff Mike Shaw, Wash, and the Old Wrangler, one long series showed Tom's effectiveness in getting juvenile delinquents to read a book called *I Dare You.* The point of the book was that any real he-man would do anything on a dare. Thus, the challenge to do good was impossible to ignore, and bad boys turned good.

According to his announcer, Tom Mix was America's favorite cowboy. In "The Mystery of the Vanishing Village," Tom Mix and Sheriff Mike Shaw investigate a village of 600 people which has disappeared. Tom's friend, Sergeant Hank Smith, searches for the girl he has come home to marry but who is missing with the village. They examine the body of the station agent who has just been killed by a murderer armed with a long range rifle armed with a silencer. Mr. Dane, a movie director, is with them, and he gets his idea for a new picture from it all. Hank visits Tom in his darkened office at the ranch as a coyote yaps.

Hank: But, Tom, I . . . I just can't help feeling about you being brought all the way from Antwerp from a hospital bed to solve a case which hasn't got any solution.

Tom Mix: Every mystery has a solution, Hank. Every question has an answer. We'll find the answer to this one.

Tom tries to solve the mystery noting that the Hollywood director, the Great Dane, had something to do with it. Suddenly Hank shouts. He sees his girlfriend Mary stumbling towards the office. They jump through the window as the show ends. Mary is hurt badly! The announcer asks—What is the meaning of this new development? See if you can figure out the mystery and tune in next Monday.

It kind of leaves you in the air, doesn't it? Well, every program attempted to leave you up in the air. Even when an episode came to an end there was always something introduced at the last moment to make you want to tune in the next day to see what happened.

Free Rings 'n' Things

Did you ever see an atomic bomb ring? In 1947, Tom Mix and Kix cereal offered them for fifteen cents and a boxtop. In a dark room you could see it sparkle. Most radio shows made similar offers.

In the 1930s, listeners to Buck Rogers could get pictures of the main characters from that series if they sent in a metal strip from a can of Cocomalt. Later, on other shows, there were offers of bombsights with toy bombs, rings with whistles, sirens, or magnifying glasses on them, secret decoders, Ovaltine shake-up mugs, a pedometer, badges, and medals.

These items were used on the programs. For example, Tom Mix used the whistle ring offered on his program to summon help from across a valley. Kids were disappointed when they found that their whistle ring could not be heard in the next room.

These special offers and gifts were an important part of old-time radio. They made kids feel as if they belonged to a program and this loyalty made them keep listening and, the sponsor hoped, buying his product.

Today, adult collectors pay high prices for these items they once had as children. For example, a *Radio Orphan Annie Decoder* (a badge which permitted you to read a message in code) is now worth about thirty-five dollars.

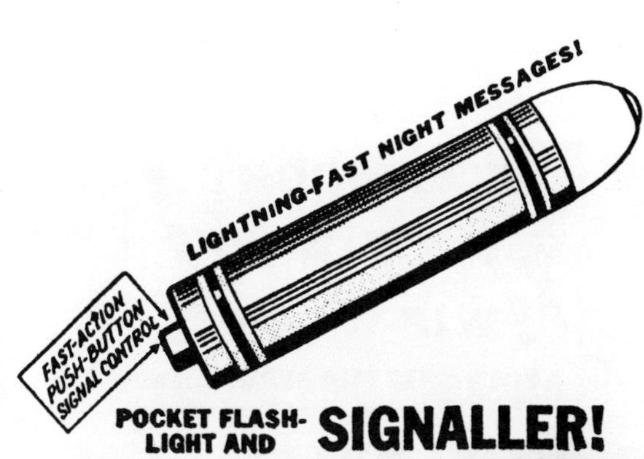

Best way to send secret night messages! Handy pocket-size flash, only 3 in. long—yet casts bright beam a long way! Easy to flash accurate messages with push-button control. Also for hiking and scouting. Bright colored metal, with silver and black bands. Comes complete with bulb and battery.

1941

1931

From Radio to TV

Do you know that many actors now on TV once played radio parts? Dick York, who plays Samantha's husband on *Bewitched,* once played both Billy Fairfield and Dickie on *Jack Armstrong.* (It was common for an actor to play two roles or more on the same show by changing his voice slightly. It saved the sponsor the cost of hiring another actor.) Bud Collyer, whom you have seen as the MC on *Beat the Clock,* once played Superman. He was also Patrick Ryan on *Terry and the Pirates.* George Gobel, now a comedian, once played the role of Jimmy, a kid, on *Tom Mix.* Agnes Moorehead was the Dragon Lady, Terry's enemy on *Terry and the Pirates.* You've seen her. She's Samantha's mother on the TV program *Bewitched.*

Longer Shows

In the 1950s, the trend was away from the fifteen-minute show to a half-hour show. These shows were complete and did not require the listener to tune in each day to find out what was going on. One popular program was *Sky King,* who, with his niece Penny and nephew Clipper fought villains like Dr. Shade.

The series began to die out the 1950s. The old favorite *Jack Armstrong* became a new program, *Armstrong of the SBI* (Scientific Bureau of Investigation). Some of the programs like *Superman* made the change to television. Others gave way to new heroes: Tom Mix to Hopalong Cassidy. By the end of the 1950s they were gone. Children watched television and ignored radio.

Jack Armstrong

Jack Armstrong, The All American Boy, an adventure serial for kids, began in 1933 and lasted until 1951.

Jack Armstrong—Charles Flynn
Billy—Dick York
Vic—Roland Butterfield (?)
Rowers—Butler Manville (?)
Sergeant—William Green (?)
Creator—Robert Hardy Andrews
Director—David Owen
Writer—Paschal N. Strong

(OPENING COMMERCIAL)

ORGAN AND VOICES: Jack Armstrong ..... Jack Armstrong ..... Jack Armstrong .....

BOB: *The All-American Boy!*

ORGAN: THEME:

BOB: Wheaties ..... "Breakfast of Champions" ..... bring you the thrilling adventures of *Jack Armstrong, the All-American Boy!*

ORGAN: THEME OUT:

BOB: Today we're saluting a baseball champion ..... a hard-hitting first baseman who's thirty-two years old today.

Who am I talking about, you ask? Well, I'm referring to Nick Etten of the New York Yankees ..... a fellow I think you'd like to know.

Nick is a quiet sort of a guy . . . . . not one to push himself forward. But at the same time, he does a terrific job. In fact, Nick is the only first baseman since Lou Gehrig's day who has held this job with the Yankees for two full seasons.

Now Nick, like so many champions, is a Wheaties eater. And here's what he says about the famous "Breakfast of Champions" . . . . .:

"When breakfast time comes around, just bring on the milk and the fruit, and the Wheaties, Breakfast of Champions, and I'm ready to go!"

So, why not take a tip from Nick Etten? Build your breakfast around big bowls of those flakes of one hundred-percent whole wheat, with plenty of milk and fruit.

Try 'em, won't you? That General Mills product—Wheaties, the "Breakfast of Champions."

(LEAD IN)

ORGAN: DRAMATIC CHORD:

ANNOUNCER: And now, Jack Armstrong in "The Adventure of the Devil's Castle."

ORGAN: MYSTERY AND SUSPENSE:

ANNOUNCER: In the laboratory of the Devil's Castle, Vic Hardy and Billy are staring at Jack with amazement. For Jack has just told them that he knows who killed the brilliant scientist, Dr. Seybold. Many startling things have happened since they found Dr. Seybold's body slumped over his desk in that lonely stone house in the Pennsylvania mountains, the house which a fanatical hermit calls the Devil's Castle. First, they found the hidden dictograph which told how the crime was committed. They discovered that Dr. Seybold had perfected an outstanding device to use cosmic rays to unlock the electrical energy of the atom, and that the murderer had stolen half of the device but couldn't find the other half. Jack and Vic found the other half, and have it well hidden. But then Dr. Seybold's body disappeared before the police arrived, and was later discovered in the hermit's cave by the lake. Dr. Seybold's lawyer, Mr. Kent, arrived, and later Professor Rowers, who offered to help Jack reconstruct the missing half of the atomic device. And then the wild hermit came up with his superstitious mountaineers in a vain effort to destroy what they thought was the devil's workshop. But right now, a bit dazed from recent events, Jack and Billy and Vic are alone in the electrical laboratory, and Jack has sprung his great surprise. Listen:

BILLY: Jack! Did you say that . . . . .

JACK: I said I know who the murderer is, Billy. And he'll be here in just a minute.

VIC: All right, Jack. Who is he?

JACK: Professor Rowers, Vic.

VIC AND BILLY: Professor Rowers!

JACK: That's right. And the lawyer, Mr. Kent, is his accomplice.

BILLY: Well, great whales and little fishes! I . . . . . I can't believe it, Jack. And here we were, talking to him just before the hermit arrived.

28 Janet Van Loon in the first national radio program for children ill and home from school.

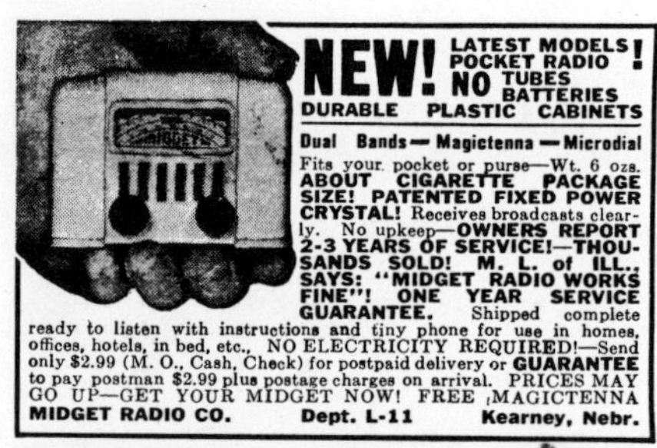

1942

1942

1941

JACK: That's right. I wasn't sure until the hermit got here. Then, when Professor Rowers wouldn't stay to meet the hermit, I knew.

VIC: Are you just guessing, Jack? Or do you know?

JACK: Let's call it an educated guess, Vic.

BILLY: Well, go ahead. Tell us how you found out. And quick . . . . . before they come down from the tower.

JACK: You remember, Billy, there was something familiar about Professor Rowers. You noticed it too. We couldn't quite place it, but . . . . .

BILLY: But it was as though we had met him before.

VIC: That's odd, Jack. I had that queer feeling too. But I didn't mention it. After all, Professor Rowers telephoned from Pittsburgh the morning after the murder.

BILLY: And I *know* I never met him before to-day.

JACK: You met part of him last night, Billy.

BILLY: *Part* of him?

JACK: Yes . . . . . his voice.

VIC: Jack . . . . . do you mean that his voice is the same voice as the one on the dictograph record?

JACK: That's right, Vic. It's the same voice we heard talking to Dr. Seybold . . . . . just before the shot was fired.

BILLY: But . . . . . but that voice was smooth and drawn out, Jack. Professor Rowers speaks with short, jerky sentences.

VIC: Of course, he might be disguising his voice now.

JACK: Not *now,* Vic. He disguised it when he killed Dr. Seybold. He had worked with Dr. Seybold before . . . . . years ago. He hadn't planned to kill Dr. Seybold at first. He wore a mask and disguised his voice so Dr. Seybold wouldn't recognize him.

VIC: Hmmmm. I do recognize a slight similarity between the two voices, Jack, as I remember them.

BILLY: But . . . . . but he came back and stole the dictograph record, Jack. He's destroyed it by this time. We can never prove it's the same voice.

VIC: Just a minute, Jack. There's something wrong with the picture. He couldn't have come back and stolen that record. We know he was in Pittsburgh about that time. We checked on the telephone call.

JACK: That's right, Vic. That's where Kent comes into the picture. *He* stole the record. He came in with the police, remember. He discovered about the record, in fact, he played it back. So he destroyed it after we went to bed.

BILLY: Then he . . . . . he was the one who tore up the study looking for the magnetic concentrator.

JACK: It *had* to be he, Billy. That's how I connect him with the case.

VIC: All right, Jack. Let's check your theory with the facts that we know. The body disappeared just before the police came. Rowers *could* have taken it away.

JACK: He took it through the underground coal mines to the hermit's cave.

VIC: All right. Maybe he wanted to throw suspicion on the hermit. After that, he would have to get back to Pittsburgh.

BILLY: He had plenty of time to do that. He didn't telephone until early this morning.

JACK: That's right. Said he saw in the papers about the murder . . . . . and wanted to help us with Dr. Seybold's invention.

VIC: Hmmm. And it was Mr. Kent who vouched for Rowers. Said he was a good friend of Dr. Seybold's.

JACK: Well, he used to be. But they parted company years ago.

BILLY: But where does the hermit come in, Jack?

JACK: That's where I'm guessing, Billy. We know that the hermit is a fanatic, but he's sincere. My guess is that Rowers has planned this thing for years. When he parted company with Dr. Seybold, he knew that the doctor might succeed. He had lived here with the doctor . . . . . somehow he discovered the old underground exit to the coal mine and the cave.

BILLY: But I don't see how the hermit . . . . .

JACK: I've found out that the hermit only moved into that cave two years ago. That probably upset Rowers' plans to get into the house through the cave from time to time. But he made the best of it. He posed as a religious man. He told the hermit that evil was loose in this house. He probably convinced the hermit that Dr. Seybold was trying to destroy the world with an atomic bomb.

VIC: Hmmm. That could be. And he hoped to throw suspicion on the hermit by taking the body to the cave.

JACK: But then he couldn't afford to let the hermit see him in this house. The hermit would know he was a fake.

BILLY: Now I get you, Jack! That's why he made some excuse and left this room . . . . . when you said you'd bring the hermit in.

JACK: That was all I needed to make sure, Billy.

VIC: I believe you're right, Jack. But we haven't a shred of evidence. If the police get hold of him, we'll never find where he's hidden that energizer. And we've got to find it.

JACK: No use even telling the police, Vic. They'd laugh at us. Rowers has a reputation as a well-known scientist. We've got to keep this to ourselves for a while.

BILLY: You mean, we've got to work with those two murderers, and pretend they're our friends, and . . . . .

JACK: That's just what I mean, Billy. It'll be a game of wits between us. We're after that energizer which Rowers has, and he's after that magnetic concentrator which we have.

VIC: Jack is right, Billy. It's going to be a dangerous game. But whoever wins will have the complete cosmo-tomic energizer.

1942

1948 **Radio Details**

P.M. Friday

8:00 WNBC: *Cities Service Band of America*—Paul Lavalle leads the 48-piece band in march music.

8:30 WNBC: *Who Said That*—Paul Gallico, Paul Douglas, George Fielding Elliot and Richard Harkness are guest experts.

8:30 WJZ: *This Is Your FBI*—Presents a case involving a paroled criminal who resumed his career of crime at the behest of a crooked politician.

8:30 WNYC: *Evenings With Great Masters*—Presents excerpts from "Romeo and Juliet," by Berlioz and "Harold in Italy," by Berlioz.

9:00 WNBC: *University Theater*—Adolphe Menjou plays the lead in "The Purloined Letter."

9:30 WNBC: *Red Skelton Show*—Red stars in satirical sequences as Clem Kadiddlehopper, Willie Lump Lump, Deadeye and Junior.

9:30 WCBS: *Musicomedy*—Johnny Desmond, Julie Conway and Kenny Bowers in "Watch Dog."

10:00 WNBC: *Life of Riley*—stars William Bendix as Chester Riley.

10:00 WOR: *Meet the Press*—a headline personality is interviewed by four newsmen.

11:15 WCBS: *Humanity's Report*—Basil O'Connor, president of the American National Red Cross, gives report on the proposals to outlaw atomic and germ warfare.

P.M. Saturday

2:00 WCBS: *Stars Over Hollywood*—Presents Jane Wyman in "Little White Lies."

4:45 WOR: *Teddy Wilson Improvises*—Offers his own treatment of "The Devil and the Deep Blue Sea," "It's The Talk of the Town," and others.

6:30 WNBC: *NBC Symphony Orchestra*—Max Reiter conducts "I Traci Amanti," "Marco Takes A Walk," and "La Boutique Fantasque."

7:00 WCBS: *The Sky Is No Limit*—Special Air Force Day program.

7:00 WNYC: *Masterwork Hour*—Tchaikowsky's "Symphony No. 3 in D Major," and Wagner's "The Flying Dutchman Overture," on records.

8:00 WOR: *Twenty Questions*—William Gaxton is guest panelist on this radio version of the old parlor game.

8:30 WNBC: *Truth or Consequences*—Ralph Edwards broadcasts from Hollywood.

8:30 WOR: *Stop Me If You've Heard This One*—Lew Lehr, Cal Tinney and Benny Rubin swap gags with Roger Bower.

9:00 WJZ: *Gang Busters*—Presents the story of how a clever gang of counterfeiters were trapped through the weakness of a gambling girl-friend.

JACK: And if they win, they can't dispose of it in this country. They'd have to do it abroad.

VIC: And some foreign country having all the electrical energy they want for nothing. We'll be reduced to a pauper country.

BILLY: Golly! We *are* playing for big stakes! Jack, I don't understand one thing. Why is the lawyer, Mr. Kent, mixed up in this with Rowers? He's got power of attorney for Dr. Seybold. He could demand that we give over the concentrator.

JACK: Right. But then he couldn't use it for himself. He's in cahoots with Rowers because they've got to get the device without anyone knowing that they have it.

VIC: We've got to plan our campaign, Jack. First of all, we've got to find out where Rowers has hidden the energizer.

JACK: Correct. And that'll be a tough job.

BILLY: Why, it might be right here in the Devil's Castle.

VIC: Or it might be anywhere in the old coal mine under the house.

JACK: And he might have taken it to Pittsburgh after he killed Dr. Seybold.

BILLY: We could hunt for it for years!

JACK: We've got to find it quickly, Billy. Or he'll beat us to the punch.

VIC: Here's what we'll do, Jack. I'll arrange to have the police sergeant come in and tell us that his men have found the energizer.

BILLY: But what good will that . . . . .

JACK: I get you, Vic. Rowers will find that hard to believe. But he'll be worried just the same.

VIC: He'll be so worried that he'll go back to where he hid it, just to make sure it's still there.

BILLY: And if he goes to Pittsburgh, we'll know he took it home.

JACK: Listen! I think I hear Rowers coming down now.

VIC: I'll go and find the sergeant, and ask him to break in on us with the news.

JACK: But don't tell him why.

VIC: I won't. We'll have to dig up more evidence before we tell him about Rowers. (GOING) I'll go now.

BILLY: Jack, I don't know if I can talk naturally to Rowers . . . when I know he's the murderer.

JACK: You've *got* to, Billy. If he suspects that we know about him, it'll make our job a lot harder.

BILLY: Yeah, and he'll bump us off at the first chance. (DOWN) Here he comes now, Jack.

JACK: (UP) Hello, Professor Rowers. Feel better now?

ROWERS: (COMING) Much better. Wonderful thing, mountain air. Hermit gone, I see.

BILLY: (SOTTO VOCE) It *is* the murderer's voice, Jack!

JACK: (WRYLY) Yes, the hermit is gone, but not forgotten.

ROWERS: Eh? How's that?

JACK: He destroyed the secret drawings.

ROWERS: Destroyed the drawings! Can't believe it!

BILLY: Well, he did, just the same. He snatched them off this table and threw them into the fireplace.

JACK: He did it so quickly we couldn't stop him.

ROWERS: Terrible. Quite a fanatic, the hermit.

JACK: So now we can't reconstruct the energizer. Unless you know enough to build one without the drawings.

ROWERS: Difficult. Very difficult. Dr. Seybold worked for years. Only one thing to do.

JACK: What's that?

ROWERS: Catch the murderer. Catch the murderer and find the energizer.

BILLY: You're right, Professor Rowers. We've certainly got to catch the murderer.

ROWERS: Could be the hermit, you know.

JACK: Yes, it could be.

ROWERS: After all, found the body in his cave.

BILLY: Do you think he's the murderer, Professor Rowers?

ROWERS: Don't know. Seems likely. Queer sort of fellow.

JACK: But he wouldn't have stolen the energizer.

ROWERS: Who knows. Wants to destroy it. Suggest we look in cave.

BILLY: (SLYLY) You wouldn't want to come with us, would you?

ROWERS: Hardly think so. Matter for police. I say, Mr. Armstrong ..... or Jack. Don't mind my calling you Jack, do you?

JACK: All my friends call me Jack.

ROWERS: Good. Proud to be your friend. I say, Jack, if you'll show me the magnetic concentrator, may help. May get some ideas for the energizer. After all, did work with Seybold, you know. Years ago.

JACK: I'd rather not just now, Professor Rowers. Too dangerous for you.

ROWERS: Dangerous?

JACK: I'll say. Don't forget the murderer is still around loose. And he may be right here in the Devil's Castle.

ROWERS: (STARTLED) Eh? How's that?

JACK: He may be hiding here in the house. After all, he seems to know this place like his own home. He'll kill to get that energizer.

RADIO FACTS!

Shows for Kids!

Other shows for kids were *Smilin' Ed* (with Froggy, the Magic Gremlin) and *Big Jon and Sparky* ("There's no school today!"). There were *Uncle Don, Uncle Remus,* and *Our Barn.* These shows were variety shows intended for small children; there was much singing, joking, and storytelling. *The Horn and Hardart Children's Hour* was a slightly different type of variety show; it sponsored children doing amateur acts (tap dancing was popular). The most famous show for young children was Nila Mack's *Let's Pretend,* which dramatized fairy tales like "Little Red Riding Hood."

1942

Listening In
With Ben Gross

Glenn Miller's band has just recorded a new tune, "The President's Birthday Ball," written by Irving Berlin especially for the infantile paralysis campaign. Proceeds from the records and sheet music will be contributed by Berlin and Miller to the March of Dimes fund. The millionth copy of Glenn's recording of "Chattanooga Choo-choo" with his autograph, will be auctioned off at the Ball in the Waldorf Astoria, Jan. 30 . . .

Glenn Miller

* * *

"Out of the Ivory Tower," a series which is designed to familiarize listeners with our great contemporary poets, bowed in over WQXR at 2:30 yesterday afternoon. Conducted by Eve Merriam, herself a young poet, the first guest was Mark Van Doren, the Pulitzer prize winner. The program opened with an interview, during which Van Doren discussed his youth. This, he followed with selections from his works. An interesting program which sets out to prove that poets are human beings who lead normal lives, work for a living and have interesting things to say.

1942

BILLY: (INNOCENTLY) And you're much too valuable to the country, Professor. We couldn't possibly risk your life until we catch the murderer.

ROWERS: Ridiculous. If I'm to help, must insist on seeing concentrator.

JACK: I promise you can see it, Professor Rowers, just as soon as we catch the murderer.

BILLY: (SOTTO VOCE) And a lot of good that'll do *you,* Professor.

ROWERS: Can't agree, Jack. Must see it now. Hate to do it, but will ask Mr. Kent. He has power of attorney.

JACK: Not power of attorney over the police, Professor Rowers. And the police are taking care of it. By the way, where *is* Mr. Kent?

ROWERS: Oh, roaming around somewhere.

BILLY: (SOTTO VOCE) Yeah, looking for the concentrator.

JACK: Here comes the sergeant now, Professor Rowers.

BILLY: And here's Vic too, Jack! Say, he looks as though he has important news!

JACK: (UP) What's up, Vic? You look pleased as Punch.

VIC: (COMING) We're getting somewhere, Jack. The energizer has been found.

ROWERS: (STARTLED)The energizer! Found? Why, why that's wonderful!

BILLY: (SOTTO VOCE) Boy! That threw the Professor for a loss!

VIC: Tell us about it, Sergeant.

SERGEANT: Not much to tell. I had my men looking, here and other places. One of them found it.

ROWERS: But I say! He wouldn't know what it looked like.

SERGEANT: We got a description of it from the drawings.

ROWERS: Where did they find it?

SERGEANT: Right where the murderer hid it. And I'm not telling anything more until we pick him up.

ROWERS: Then, then you don't know who he is yet?

SERGEANT: Not yet. But we'll find him.

ROWERS: Splendid. Jack, when you get the energizer, let me know. I can help set it up.

JACK: But aren't you staying here until we find the murderer, Professor?

ROWERS: Waste of time. Matter for police. Go back to Pittsburgh until you're ready for experiment.

BILLY: Aw, don't go back to Pittsburgh, Professor. Wait till we catch the murderer.

ROWERS: Sorry. Important matter to attend to. Take Mr. Kent with me. He wants to get back, settle the estate, you know. (GOING) But I'll come back soon.

SOUND: DOOR CLOSING:

JACK: (DOWN) Boy! He couldn't wait to get away!

SERGEANT: Now look here, Mr. Hardy. I did what you asked me to. What's it all about?

VIC: We'll tell you later, Sergeant. It's nothing you can act on now.

SERGEANT: (DOUBTFULLY) Well, if it weren't for the reputation you and Mr. Armstrong have . . . . .

JACK: When we can prove who the murderer is, Sergeant, we'll tell you at once.

SERGEANT: All right. I'm going to the hermit's cave. I've got some questions to ask that guy. (GOING) He knows something he won't tell.

VIC: We're getting places, Jack. We know that the energizer is in Pittsburgh.

JACK: Right. Rowers is going back to check on it. We'll follow him in and see where he's hidden it.

VIC: And I'm going to be armed, Jack. If he finds us in there, he'll know we're on to him.

JACK: I'll say. He'll have to kill us then. Come on, Billy, let's get our things. We're going to march right into the lion's den!

MUSIC: ORGAN: DRAMATIC TAG:

(LEAD OUT)

ANNOUNCER: Right into the lion's den! And you can bet that there is going to be plenty of fireworks in that particular den when Jack and his friends arrive. You won't want to miss it, so listen in, all of you, to the next thrilling episode of "The Adventure of the Devil's Castle," with *Jack Armstrong, the All-American Boy!*

(CLOSING COMMERCIAL)

JIM: This is Jim Butterfield, and I'm wondering . . . . . are you satisfied with the progress you're making in sports? Are you sold on the game of football or baseball you're playing? Or do you figure, with some expert coaching, you could play a heck of a lot better?

Well, look. If you want coaching, I'd advise you to find out about the Wheaties Library of Sports. These seventeen books are especially written for kids like us. They're full of action pictures, and hot coaching tips on almost every major sport, tips that will help you play a better game.

It's a swell deal, this Wheaties Sports Library. What's more, it ties right in with the Breakfast of Champions idea . . . . . that eating right, and the right kind of exercise, are mighty important in helping you and me grow up strong and healthy.

Now, eating right means putting away three good meals a day, starting with a nourishing breakfast. Say, one built around big bowls of those whole wheat flakes, Wheaties, with lots of milk and fruit.

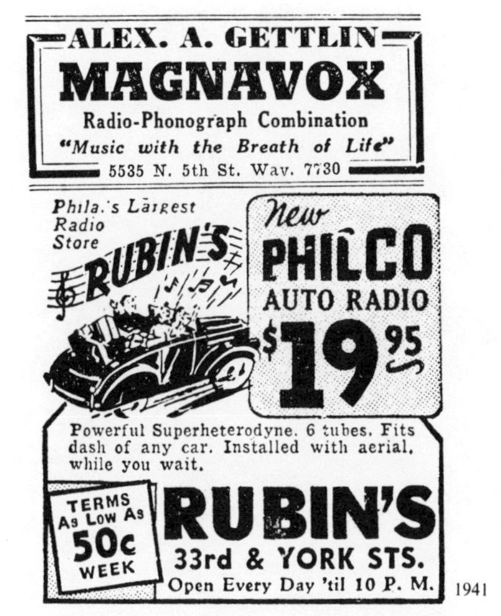

1941

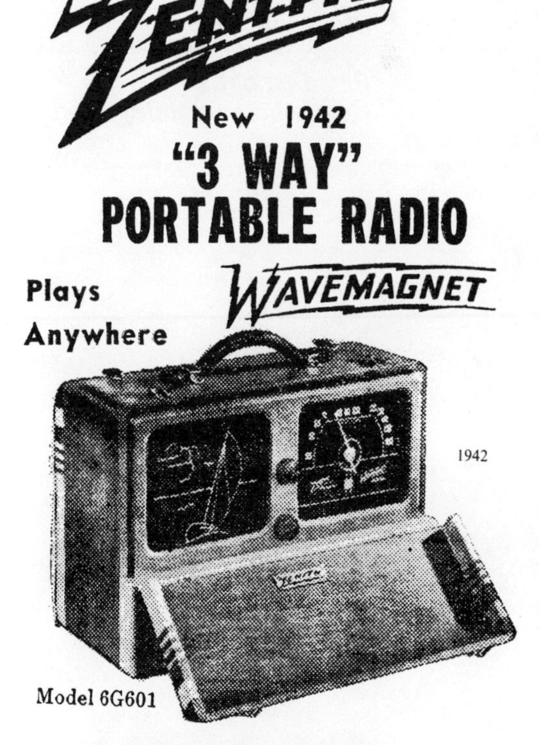

1942

As for the right kind of exercise . . . . . well, you'll read about the Wheaties Library of Sports on your orange and blue package of Wheaties . . . . . Breakfast of Champions.

Why not check up on the Wheaties Sports Library today?

MUSIC: THEME:

JIM: This is Jim Butterfield . . . . .

ANNOUNCER: (CUTS IN FAST) and Bob McKee . . . . .

JIM: for General Mills, makers of Wheaties, Breakfast of Champions . . . . . who invite you to listen again tomorrow to another exciting episode of *Jack Armstrong, the All-American Boy!*

(Whistle)
Rinso White.
Rinso Bright.
Happy little washday song.

1941

Chapter 5

Boo hoo! Housewives, while ironing, wiped tears from their eyes as they listened to how others mismanaged their lives in daily fifteen-minute-long radio serials called soap operas (most were sponsored by soap companies).

Real People?

Soap operas seemed to be about real people. For example, Betty and Bob ran *The Trumpet* (a newspaper), and *The Carters of Elm Street, The Couple Next Door,* and *Vic and Sade* seemed like married couples who lived in the listener's neighborhood.

There were serials about men like *Just Plain Bill, David Harum,* and *Lorenzo Jones,* but the serials seldom drew a male audience. Most popular were those shows that revolved around a female character like the ever-suffering lead in *Stella Dallas* or the women in *The Romance of Helen Trent, Myrt and Marge, Big Sister* or *Ma Perkins.*

Soap Opera Plots and Themes

Usually soap operas dealt with emotional situations, unappreciative children as in *Stella Dallas* and the search for romantic love as in *Helen Trent. Mary Noble, Backstage Wife* gave the average woman some insight into how it might feel to be the wife of a famous Broadway actor. At one point, a backer of her husband Larry's plays has become so enamoured of Mary Noble that he hires a woman to make Larry fall in love with yer.

Love (and its complications) was predominant. In *Our Gal Sunday,* for example, Joyce Irwin, Lewis Carter's ward, finds he is romantically attracted to her. She either leaps or falls from a window and is killed. In addition, there were the newly appearing long-lost relatives, and probably the most overworked device of all, the amnesiac. Crime, illness, and economic troubles were also common (accounting for about one-third of program time).

Reasons for Popularity

What was it in those shows that captured the attention of women during the long afternoons in Iowa or Maryland as well as Massachusetts? Why did they follow programs which went on and on and on? To be sure, the daily segments seemed like visits from old acquaintances. To a relatively naive audience, those radio voices right in one's own kitchen sounded like close friends spilling out their troubles. Soap operas offered advice, said 41% ot the listeners in a 1942 survey.

Too, the soap opera served as escapism, as a form of electronic gossip—women could always converse about the plight of their common friend, Stella Dallas. You see, poverty-stricken Stella (whose daughter Lottie wed wealthy Richard Grosvenor) claimed it would be best if she stayed out of Lottie's life forever. However, she didn't. And daily, housewives wept together over her plight.

Since the plots never came to a conclusion without starting something new, the audience was hooked on following them daily without missing an episode. In a way, this meant that the listener belonged; she was in on what was really happening. Curiously, radio became reality instead of fantasy—it was not clear if the real world was the one in which the listener lived or the one which she tuned in. Soap opera characters solved complicated problems; real people often cannot do so.

29 In 1935, the *David Harum* program. — (NBC photo)

30 Virginia Clark played the title role in the serial *The Romance of Helen Trent,* heard over the WABC — Columbia network Mondays through Fridays from 12:30 to 12:45 p.m. (EST). — (CBS photo)

31 The *Stella Dallas* cast in 1939. — (NBC photo)

32 Alan Bunce played the young Dr. Malone and Elizabeth Reller played his wife, Ann Richards Malone on *Young Dr. Malone,* a popular soap opera about a young physician at Three Oaks Medical Center. — (CBS photo)

The characters and the commercials helped to create the American myth. The commercials played as much a part in forming the ideals of American women as the soap operas' story lines and characters. A housewife protected her family's health, got their clothes cleaner, and made their life more enjoyable if she used the products advertised on the show. She didn't if she used off-brand or unknown products, said the manufacturer.

Working class people were generally excluded from the stories; they were either skilled workers or professional people. For example, Ruth Wayne of *Rinso's Big Sister* was the wife of successful Dr. John Wayne, who with Dr. Reed Bannister (his friend and rival) ran the Glen Falls Health Center. The central character in *Front Page Farrell* was the star reporter for *The* New York Eagle. Bill Davidson of *Just Plain Bill* was Martville's barber; Perry Mason, a lawyer; Ma Perkins, owner of a lumberyard; and Pepper Young of *Pepper Young's Family* was Mayor of Elmwood.

Early Soap Operas

One of the earliest soap operas was *The Career of Alice Blair* (1935). Another of these early fifteen-minute programs was *Clara, Lu 'n' Em,* billed as "a backyard gabfest sponsored by Super-Suds." Also popular in 1934 was *The Romance of Helen Trent*. This program, which continued into the fifties, dealt with the loves of Hollywood designer Helen Trent who at one time was accused of murdering a producer (although he was really murdered by a friend of her enemy, Cynthia Swanson).

Longest Running Soap Operas

One Man's Family was one of the all-time greats. It was on the air from Friday, April 29, 1932, through May 8, 1959, totalling 3,256 programs. Its creator was one of the most famous names in radio, Carlton E. Morse, who also directed most of the shows. It began as a half-hour program over the Pacific network of NBC and began to catch on when others listening at night on their sets began DX-ing or picking up distant signals (radio signals travel farther at night because of more pleasant atmospheric conditions). Thus, people began writing and asking for local broadcasts of the show. The network began a weekly transcontinental program from San Francisco every Saturday night at 8:00 PST over thirty-eight stations.

According to one report in a contemporary magazine, *Radio Stars*, the new audience didn't know the beginning of the story, so the network felt obliged to begin all over again on its Saturday night broadcasts while the Wednesday night broadcasts continued as before.

Other Shows

Just as *One Man's Family* was a typical soap opera of the thirties, *When a Girl Marries* was typical of the forties. It went on the air on May 29, 1939, written by Elaine Carrington and Leroy Bailey. Its star, Mary Jane Higby, remained on the show for eighteen years. The show began with a musical interlude and the announcer spoke:

> *When a Girl Marries.* This tender, human story of young married love is dedicated to everyone who has ever been in love.

The central character, Joan Davis, had a husband Harry who while suffering from amnesia became engaged to another woman. Harry's amnesia lasted many years.

33

34

35

36

33 *Portia* (Lucille Wall) *Faces Life* in 1941. Sound effects man at left. — (NBC photo)

34 *Pepper Young's Family,* Burt Brazier and Betty Wragge. — (NBC photo)

35 Claire Neisen and Jimmy Meighan in *Backstage Wife.* — (NBC photo)

36 *When a Girl Marries* in 1945 (l. to r., Georgia Burke, Mary Jane Higby, Robert Hoag, Dolores Gillen). — (NBC photo)

That title was evocative. Before the days of Women's Lib, the chief ambition of most women was to get married. Even though they might dream of becoming a doctor or lawyer, society dictated that they be more interested in being a wife and mother. However, they could escape their humdrum household chores whenever they wished if they listened to lawyer Portia Manning in *Portia Faces Life,* or *The Life and Love of Dr. Susan.*

The Audience

The main reason soap operas did not appeal to children is that they lacked action—no smashing, banging, or fistcuffs—and further, they used themes like love which young children found uninteresting. The people on the show always seemed to be middle-aged and worried about the kinds of problems that were boring to children. *Amanda of Honeymoon Hill*, for example, heard over NBC Blue in the forties, dealt with love and marriage in "The Romantic South." In addition, the voices of some of the actresses and actors seemed unappealing and unlikely to capture the interest of children. Stella Dallas had a tired worn voice that probably sounded too much like one's own mother. Kids generally were bored with the serials, but when confined to the house with a cold they generally listened to them for want of something else to do.

Of course, men were usually working during the daytime and the daily serials held little interest for them. Men's radio was usually thought of as news and commentaries with perhaps a humor or mystery show as a diversion. There were no men's shows that one could listen to during the daytime.

Women listened because they could lighten the drudgery of their daily tasks, those which could be performed mindlessly while they followed in their imaginations the activities of others. They need not watch a TV screen (as women do today) or stop to read.

The Sponsors

The sponsors were usually manufacturers of soap, like Super-Suds, Oxydol, and Rinso. By and large, most soaps were alike and the ads served only to switch users from one product to another, not to create new markets. That is, the total amount of soap sold remained the same even if it was not advertised, but the amount of each manufacturer's share of that total sales changed.

The manufacturer got his money's worth because soap operas were very cheap to produce. The most expensive radio programs were musical shows; the dramas required only a few actors and frequently one actor did several voices. That is, he might use a low voice for one character, a dialect for another, and a high voice for a young child. Of course, the actor had to be careful not to mix them up although this did happen on occasion. For example, on one show, an actor spoke the child's lines in a deep gruff voice.

Frank and Ann Hummert

By 1938, one writer counted 78 soap operas on the air. It was a busy industry and perhaps even a new art form. Frank and Ann Hummert are often credited with creating it. They created *Betty and Bob* as well as the famous *Just Plain Bill, The Romance of Helen Trent, Lorenzo Jones, Mary Noble, Backstage Wife* and *Young Widder Brown.* They operated as a creative team to indicate plot lines, then had writers fill in the skeleton with the dialogue and sound effects. This technique (also used by such writers as Charles Dickens) permitted them to turn out enormous numbers of scripts.

The Hummerts amassed a fortune from soap operas, earning large fees in a period before taxes and inflation ate up money faster than you could make it.

According to Mary Jane Higby, an actress on many of their shows, the Hummerts were careful with the production of their shows as well. As she points out, effects or music were not permitted to "obscure one word of the dialogue." Air Features (Hummerts' company) insisted that the actors read dialogue clearly and distinctly.

The Soap Opera Formula

The beginning of a soap opera always included a lengthy recap. Since very often little happened over a long period of time, the device was not really necessary for regular listeners. Further, it shortened the daily fifteen-minute serial (with commercials) to about ten minutes of actual drama.

Writers always strived to make scripts seem like real life and at least one show used an ingenious device to simulate reality. Wendy Warren was news reporter for the first part of her daily radio show. After a commercial, the remainder of the program was a typical soap opera. On one lengthy series, Mark Douglas hopes to marry Wendy after he gets a divorce but Wendy decides to give him up when his wife pleads with her to do so. Thus, fact and fiction were blended.

Writers were adept at setting scenes fast. They did so by using a familiar locale (a train station, for example, would be identified by sounds of a steam engine or by having the announcer set the location verbally). Once the scene was set, the sound would fade out and come back in at odd moments so that the listener wouldn't forget the location. As in real life, we tend to block out the irrelevant sounds. Try it. Stop and listen as you read the book. Do you hear sounds that you had not noticed before? Keep listening and you'll hear more and more sounds that you blocked out as you were reading the words. This phenomenon was exploited by the writers, so that if the characters were inside a boiler factory, sound effects men did not have to recreate all the noise continuously. (In such a situation, however, one character might ask the other to walk into a quiet office.)

Frequently, it was necessary to remind the listeners who the characters were, especially if there seemed to be many characters and many plots and sub-plots going on at the same time. Thus, the character would often use an adjective before the name of the person to whom he was speaking. Or he would recap some of his reasons for speaking.

Often the voices of the characters were clearly identified by a kind of brief summary given the person as he spoke. "This is John, your long lost nephew, remember me?"

Generally, the soap operas were simple and made no use of complicated sound devices; most of them took place inside or in relatively quiet places. The mystery and drama shows, on the other hand, often used weird and fantastic sounds because of their setting.

RADIO FACTS!

Women on Radio!

Women performed as actresses and singers on almost every show, but some women had their own shows. Too often they are forgotten. As for singers, many had their own shows: *The Kate Smith Hour, The Andrews Sisters, The Mildred Bailey Show, The Connie Boswell Show, The Dinah Shore Show.* Women broadcast news (*Pauline Fredericks, Sigrid Schultz Reporting from Berlin, Wendy Warren and The News, Meet the Press* with Martha Rountree, *Brenda Adams/Women in the News*) as well as Hollywood gossip (*Voice of Broadway* with Dorothy Kilgallen, *Hedda Hopper's Diary*). There were women with interview shows of their own (Mary Margaret McBride, *The Nellie Revell Show, The Jane Cowl Show*) in addition to women who offered advice (Jane Porterfield on *Do You Need Advice?*). These were in addition to women who shared billing and co-starred on shows like Ed and Pegeen Fitzgerald, or Tex and Jinx. They filled other important roles as commentators, preachers, comediennes, reporters, and teachers.

Characterization

It was important that the characters remained the same. They could not do things out of character. Nora, the nurse, in *Nora Drake* tries to help Charles Dobbs, a lawyer, rebuild his life. In *Young Doctor Malone,* Dr. Jerry Malone is awed by his new job and doesn't see that Lucia Standish is ruining his life. Most serials ended each program with a question that would make the listener want to tune in to hear the answer the next day. The question on Friday would have to be the toughest of all so that the listener would want to tune in the following Monday.

Putting the Show Together

As one writer (Erik Barnouw) points out, the writer was expected to write "the equivalent of a three-act play every two weeks." Paul Rhymer, who did *Vic and Sade,* wrote a total of 3,500 scripts in his years with the program. (*Vic and Sade,* however, was not, strictly speaking, a soap opera or a serial—each day's drama was complete and self-contained which makes Rhymer's achievement all the more spectacular.)

A typical radio series cost about $1,400 to $2,000 a week with budgets arranged for a total of twenty-five actors per week or five a day. The actors were expected to rehearse about one or two hours per show. Sometimes actors got jobs in several serials and raced from one studio to another to be able to cover them. In some cases, they had a "stand-in" do the lines during rehearsal and then they came on to read the lines on the air without rehearsing.

Rehearsals were needed. The reading aloud of a script often revealed awkward and unnatural phrasings, the use of difficult-to-pronounce words (try *ridiculous*), and the cast had to iron these problems out. In a typical rehearsal, the director timed the script, cued the sound effects and added to or cut the script to make it fit the time slot. He figured out the time required for musical interludes (called bridges) to even out the show. If the timing varied during the actual broadcast, he instructed the organist to fill in with extra music or to cut a selection short.

Although at one time programs used live orchestras, a single organist could create all kinds of sounds to play on the emotions of the listening audience. Since one musician was eighteen times cheaper than an orchestra, producers and directors loved organists and their music.

The End—Thanks to TV

Radio soap operas ended when television became available to the majority of homes. However, some radio characters made the transition to television serials. The first TV soap operas were unsuccessful. In 1947, the first television soap opera appeared briefly on the Dumont Network (It was *A Woman to Remember*). Then in 1949, *One Man's Family* came to TV; in 1954 *Portia Faces Life;* and in the 1960's *Young Dr. Malone*. After 1961, soap operas existed only on television.

37 Ma Perkins

38 *Young Widder Brown* (Wendy Drew). —(NBC photo)

39 In 1937, the *Guiding Light* cast. —(NBC photo)

Ma Perkins

This popular soap opera (sponsored by Oxydol) was broadcast in daily fifteen minute segments.

Ma Perkins—Virginia Payne
Shuffle Shober—Charles Egleston
Gladys Pendleton—Patricia Dunlap
Fay Perkins Henderson—Rita Ascot?
Writer—Lester Huntley?
Director—Edwin Wolfe?
Announcer—Dick Wells

1942

MUSIC.—*Opening theme for 15 seconds.*

ANNOUNCER.—And now for Ma Perkins . . . brought to you by Procter and Gamble, the makers of Oxydol.
(Opening commercial)

ANNOUNCER.—And now here she is . . . Ma Perkins! Well, yesterday the folks heard Paul Henderson's last will and testament. It was a very short document, leaving everything to Fay. Paul's estate was a good one; so Fay's security and that of the child she is going to have are assured. But Fay doesn't care about that! Nor does she seem to care about the baby she's expecting . . . All Fay thinks about is Paul! *(low)* Can Ma bring Fay through her grief and despair . . . make Fay see that life goes on, whether we will it or not? *(Tiny pause).* Well, right now Ma and Shuffle are walking home from the lumberyard through the October dusk. Rushville Center is peaceful and quiet as . . . listen!

SHUFFLE.—Tarnation, there's a real nip in the air tonight. Another couple of weeks and we'll be taking out the old ear muffs.

MA.—Oh, we'll have Indian summer first. This is the first real chilly day we've had since we got back.

SOUND.—*Jumping cue slightly . . . an automobile horn honks, not too intimate . . . the automobile is going fast . . . second honk fades . . . the sound of the car is barely audible.*

SHUFFLE.—Oh, there's Lonnie Konvalinka in his new car . . . *(calls)* . . . evening, Lonnie!

MA.—*(Calling)* Hello, Mr. Konvalinka! *(A tiny pause . . . murmuring) She* sent Fay the sweetest note.

SHUFFLE.—*Mrs.* Lonnie?

MA.—Yes. She's *such* a nice person.

SHUFFLE.—Practically everybody in town must have sent Fay a note.

MA.—*(A sigh)* Yes. *(A tiny pause)*

SHUFFLE.—*(Very tenderly)* Come on now, Ma. It's a terrible thing, but you ain't going to let it color your whole life. You ain't going to let it discourage you.

MA.—It's going to color *her* whole life . . . Fay's. I guess on the day I die, Shuffle, I'll look back and I'll see how my life sort of divided itself up into three parts. The first was with Pa, till he . . . left us. The second part, with the children . . . getting Evey married, and Fay. And the third part . . . now. Fay being widowed, after less than a year of marriage. *(Dreamily)* Fay. Who'd ever thought that she'd be the one? The day she was graduated from grade school . . . all the little girls in their nice clean dresses . . . the day they was graduated from high school . . . why, she was the second youngest in her class, and now she's one who . . . *(Practically a throw-away)* It's funny, it's funny.

SHUFFLE.—She'll come out of it, Ma. She'll get back into the scheme of things down here. Her friends . . . taking things easy . . . she'll be okay. Remember she's young, Ma, and youth ain't *never* sat down and cried and cried and never stopped crying. In a couple of weeks you'll be taking her down to buy some new dresses, and she'll look at herself in the mirror, and just by accident her eye'll fall on a new hat, and . . . say, you name one woman who can resist the temptation of a new hat, 'specially Fay. She always did have a weakness for the darnedest bonnets I ever saw!

SHORT-WAVE STATIONS

This Afternoon and Tonight

Calls.	Location.	E. D. T. P. M.	Meg.
RKI,	Moscow	noon- 7.00	15.08
GSF,	London	noon-12.30	15.14
GSD,	London	noon-12.30	11.75
DJD,	Berlin	noon-2 A.M.	11.77
DJB,	Berlin	noon-2 A.M.	15.12
	Rome	1.30- 2.30	15.31
RNE,	Moscow	4.00-10.00	12.02
	Sydney, Australia	5.00- 6.00	11.83
EAQ,	Madrid	5.00- 9.00	9.86
LRX,	Buenos Aires	5.00-11.30	9.65
PSH,	Rio de Janeiro	6.00- 9.00	10.29
GSC,	London	7.20-12.30	9.59
RAN,	Moscow	8.00-11.00	9.62
DXD,	Berlin	8.00- 2.00	10.58
	Rome	9.00-12.00	9.76
Tomorrow Morning			
GSC,	London	1 A.M.-3.00	9.58
VLR,	Melbourne	5.00- 8.00	9.59
DJB,	Berlin	5.00-1 P.M.	15.12
GSF,	London	6.00-1 P.M.	15.14
JZJ,	Tokio	7.00- 9.00	11.80
2RO-6,	Rome	8.00- 9.00	15.31

1941

MA.—*(Half laughing but a chiding note)* Well, Shuffle, I hope you're right. But you mustn't forget that we ain't *got* too long to wait while she recovers. *(Very sadly)* If she keeps on grieving and mourning through the next few weeks . . . she and her baby both will have a . . . well, Tom Stevens is worried already. *(Almost to herself, slowly)* No . . . the thing that's going to cure Fay is . . . is . . . *this. (A tiny pause)*

SHUFFLE.—*(Mildly surprised)* Eh? What are you waving your hand at . . . what do you mean, "this" . . . you mean old man Johnson raking up the leaves there on his front lawn? *(Calls without pause)* Evening, Mr. Johnson! *(A tiny pause . . . murmuring with a smile)* Getting deafer 'n' ever, ain't he?

MA.—*(Hint of a laugh but an earnest note)* No, I don't mean Mr. Johnson 'specially, but . . . yes, Mr. Johnson among other things. *(Sincerely)* Shuffle, if we'll only look around us, we'll see so much to . . . to take the sting out of our sorrows! That's what I meant when I waved my hand at *Rushville Center.* At Mr. Johnson raking his leaves. And the smell of the October leaves being burned on 20 lawns and the yellow house lights blinking on as folks like us walk home after a day's work. Living . . . I guess what I'm talking about is living. Taking the days as they come . . . the seasons . . . living for each day itself . . . just living! Putting up the screens in May and taking 'em down in September . . . doing your work, listening on an October night to the wild geese, as a mile over our heads they go on their wonderful and mysterious journey!

SHUFFLE.—Yep . . . that sure *is* a wonderful sound.

MA.—You know, Shuffle, when I was a little girl, my father used to stand with me outside our house, of an October afternoon, and show me the wild birds going south. Looked sort of like a smoky smudge. And one year, I must have been six or so, a gray goose feather fell right at my feet. And my father laughed and he said "hold on to that, young lady the bird'll be back in the spring to get it, or maybe to drop you another feather!" And I asked my father . . . somehow it impressed me . . . "Year after year, will that same goose be flying right over our house?" He smiled sadly, and said, "If *you'll* be here to find the feather, the goose will drop it for you." *(A tiny pause)* I'm a woman grown, but I've never forgotten that little incident. And ever since I've *liked* the idea of year after year

. . . the regularity of the seasons . . . the mysterious way of God, moving those birds across a thousand miles a day and night and empty air, and me standing there, a part of it, because I . . . well, because I'm a part of it. And that's what I'd like my children to to know . . . especially Fay . . . I'd like *her* to see that if we'll only be there to find it, the gray goose feather will always come. Telling us that the world goes on, that all's right with the world. *(A tiny pause)*

SHUFFLE.—*(Quietly; he's deeply moved)* I guess that's the story of our lives, Ma . . . the lives of you and me and the rest of us who stay in all the forgotten little villages, and let the rest of the world go by. Except . . . *we* don't let the world go by . . . it's the folks in a hurry who let it by. Us, we got time to take it in.

MA.—*(Not much volume but very earnest)* Yes, Shuffle . . . that's it exactly! And that's the secret of peace. Let each day come . . . take it as it comes . . . take it for everything it has . . . and when it goes, you've lived that day! Now if Fay will only see that . . . *(A little catch . . . hint of a half wistful laugh)* Yes, if Fay will only see it.

SHUFFLE.—*(Deep breath which comes out as a sigh)* Yep, yep, yep. But Fay couldn't be in better hands, Ma, so don't worry. You got a good recipe for living, and if anybody can teach it to her its you. Just give her enough time and she'll come through with flying colors.

MA.—*(A sigh)* Yes . . . if there'll be time enough. Well . . . here we are. You said you're going some place for supper tonight, Shuffle?

SHUFFLE.—I'm having a quick bite. Then I'm going to a meeting of the volunteer fire department . . . I missed the last meeting . . . we're having a big discussion should we have a bingo party for our annual blowout, or should we do like they do in Three Rivers and give a masquerade and carnival. Opinion is divided fifty fifty, and the arguments will be coming thick and fast . . . Wouldn't miss it even for one of *your* suppers, Ma.

MA.—*(Snickering)* And on which side are you going to throw the weight of *your* opinion, Shuffle?

SHUFFLE.—Oh, whichever side needs a feller with a good loud voice . . . I'm their man. *(They laugh)* Well, good night, Ma . . .

MA.—Good night, Shuffle . . . see you tomorrow . . .

SHUFFLE.—*(Fading)* Yes, Ma. Night! And remember what I said . . . don't worry!

MA.—*(Half calling)* I won't! And I think a masquerade would be more fun than bingo!

SHUFFLE.—*(Fading, calling back)* Then I'm on the side of the masquerade, and the bingo fellers ain't got a chance!

MA.—(*Laughing to herself, fondly, but throw away*) Shuffle, Shuffle.

SOUND.—*Her footsteps cross wooden porch . . . door opens and closes on*

GLADYS.—(*Off and fading in . . . take cue from door opening)* Yes, I can see that. Did somebody come in, Fay?

MA.—(*Surprised*) Who on earth . . .

RADIO FACTS!

Soap Opera Introductions?

A story of an Iowa stenographer who fell in love with and married Broadway matinee idol Larry Noble.
Backstage Wife

This story asks the question, Can this girl from a little mining town in the West find happiness as the wife of a wealthy and titled Englishman?
Our Gal Sunday

The story of a woman who sets out to prove — that romance can live on after thirty-five.
The Romance of Helen Trent

1950

FAY.—(*Fading in; quietly; neither happily nor unhappily*) Hello, Ma. Come in. Gladys Pendleton came over to pay us a call . . . We were waiting for you.

MA.—(*Astounded, but her pleasure grows through the speech*) Why . . . why, Gladys Pendleton! How nice of you, Gladys . . . I seen your mother downtown a couple hours ago . . . it was good of you to come, child!

GLADYS.—Good evening, Mrs. Perkins. I . . . I . . . I came over to tell Fay how I . . . well, I told her that she has my sympathy. I'm sorry. It was a terrible thing.

MA.—(*Gently*) Thank you, Gladys. Fay, wouldn't Gladys like some tea . . . or maybe she'd like . . .

GLADYS.—No. thank you . . . Fay's already offered me . . . I can't stay.

MA.—Well, you'll stay till I go up and wash my face and come down again. (*Fading*) We ain't seen you in a long time, Gladys. How've you been? You're looking real well!

GLADYS.—(*Raising voice very slightly*) Thank you . . . I've been fine. *(A little pause, half sulky, half nice)* Your sister told my mother this morning that Paul left a good estate. Are you planning on traveling?

FAY.—*(Quietly)* No. I'm going to have a baby, you know . . . I really haven't thought much beyond that.

GLADYS.—I'll tell you something, Fay. Don't let money matter too much. I know. It doesn't buy the things you want.

FAY.—*(Quietly)* I know that, Gladys.

GLADYS.—*(Half defiantly, half bitterly)* I was the only kid in our class who had a fur coat. I was the only girl in high school with a roadster. So what? When I met somebody I really liked I couldn't even keep him . . . he went and married somebody else.

FAY.—*(A bit flustered)* Gladys, you . . . you mean . . . Paul? But that was so long ago . . . I . . .

GLADYS.—(*A little burst*) Why do you think I came over here? Do you think I call on everyone I happen to have a nodding acquaintance with? I came because Paul . . . because I . . .

FAY.—(*Very gently but a bit breathless, a tone of wonder*) I guess I understand. Thanks . . . thanks for coming, Gladys. That's sort of . . . it's *very* nice of you. No, it's more than that . . .

GLADYS.—Oh, it's . . . well I don't . . . (*more rapidly, even more jerkily*) I think I'll be going now . . . If there's anything I can do for you let me know, Fay . . . Huh, that's a laugh! Maybe you're the one who should be doing something for me!

FAY.—I . . . I don't understand, No, don't go . . .

GLADYS.—I must. What don't you understand? Don't you know what I'm trying to tell you? Since I'm in the mood for confession . . . I'd trade places with you in a minute . . . You're better off than I can ever be. *(Fading without pause)* Don't bother to get up . . . in a day or two maybe we can have lunch together . . . Express my apologies to your mother . . . good night!

FAY.—But Gladys . . .

SOUND.—*Door opens and closes a bit off.*

MA.—*(Brightly . . . off and fading in)* Fay, did you say that you . . . *(Stops, surprised)* Where is she?

FAY.—*(Throw away, barely audible)* Gone. *(Marveling . . . to herself slowly)* She . . . she said that she'd . . . that she'd . . . she'd trade places with me in one minute. *(A little pause)*

MA.—*(Slowly)* Gladys said that? That . . . that was wise and good of her.

FAY.—*(To herself)* She must have meant . . . the baby . . . Even a few months with Paul being better than . . . than nothing. *(Little stronger tone of wonder)* She . . . why, she must have loved Paul all along.

MA.—*(Barely audible)* Yes. Maybe. *(A little pause)*

FAY.—*(For curtain)* I . . . I guess you're right, Ma. Maybe there are other people in the world besides me.

ANNOUNCER.—To which May might say, thankfully, "Yes, it's as I said. Live in the present and the future, Fay, because life does go on and we with it." Has Fay begun to live again . . . begun to feel herself a part of this world of ours again? Tomorrow Fay and Ma have some interesting visitors . . . old friends of ours . . . so be sure to listen again . . . tomorrow!

(Commercial)

ANNOUNCER.—So Gladys Pendleton says she'd change places with Fay in one minute, and Fay sees that perhaps she's not the *only* person in the world. Well, tomorrow Fay hears more advice; and soon, we're going to hear more about Paul's will, so be *sure* to listen again!

This is Dick Wells speaking for the makers of Oxydol.

Radio

RADIO WILL COVER UNO; JACK+FRED=LAUGHS;

·MONDAY'S FEATURES·

8:00 A. M.—WEAF. WABC. News from Abroad. (Also **WNYC** at 10:30; **WNYC** at 5:45 P. M.; **WABC** at 6:45; **WEAF** at 7:15; **WNYC** at 8; **WJZ** at 11.)
11:00 A. M.—WEAF. Fred Waring Show.
11:00 A. M.—WJZ. Breakfast In Hollywood.
2:15 P. M.—WOR. Opening of UNO Security Council. (Also **WEAF, WJZ, WABC, WHOM, WQXR, WEVD, WNYC, WNEW, WHN, WMCA** at 2:30, Sec. of State Byrnes, Gov. Dewey, Mayor O'Dwyer.) (Also Resumes and Recorded Highlights over **WMCA** at 4:45 and 8:30).
5:00 P. M.—WABC. American School of the Air.
6:30 P. M.—WABC. Alger G. Chapman, "Your State Income Tax."
7:30 P. M.—WABC. Bob Hawk Show.
8:00 P. M.—WEAF. Cavalcade; Agnes Moorehead in "Whither Thou Goest."
8:00 P. M.—WABC. Vox Pop.
8:00 P. M.—WHN. Author Meets Critics; Morris Ernst.
8:30 P. M.—WEAF. Jussi Bjoerling, Barlow's Orch.
8:30 P. M.—WJZ. Dashiell Hammett's "Fat Man."
8:30 P. M.—WABC. Joan Davis Show.
9:00 P. M.—WEAF. Jascha Heifetz, Voorhees' Orch.
9:00 P. M.—WJZ. William Gargan in "I Deal in Crime."
9:00 P. M.—WABC. Radio Theatre; Danny Kaye, Virginia Mayo in "Wonder Man."
9:30 P. M.—WEAF. Information Please; Thomas Costian, Gladys Schmitt.
10:00 P. M.—WHN. Giovanelli-Lewis Fight; Steve Ellis.
10:00 P. M.—WEAF. Andres Segovia, Barry Wood, Faith's Orch.
10:00 P. M.—WABC. Screen Guild Players; Jeanette MacDonald, Nelson Eddy in "Sweethearts."
10:15 P. M.—WOR. Secreet-Walker Fight; Sam Taub.
10:30 P. M.—WABC. Helen Hayes "Life of St. Frances Xavier Cabrini."
10:30 P. M.—WJZ. Question for America; "Should Our Atomic Bomb Secrets Be Revealed to the UNO?" Dr. Harrison Brown, Sen. Joseph Ball.
10:45 P. M.—WMCA. The UNO This Week; Benjamin V. Cohen (Premiere).
11:30 P. M.—WEAF. Eileen Farrell Show; Earl Wrightson.

1946

Super Suds, Super Suds,
Lots more suds with Super Suds.
Richer, longer lasting, too.
They're the suds with super-do.

Chapter 6

Want to get rich quick? Get on a quiz show. Many people did, for quiz shows zoomed to popularity. Why? They were good entertainment, had some educational value, and they gave away money, refrigerators, and cars. No wonder quiz shows were smash hits.

Quiz Kids

Some quiz shows used experts; listeners tried to stump them. If they did, the sender of that tough question won a prize. One of the most popular panel shows was *The Quiz Kids,* a program from Chicago, sponsored by Alka-Seltzer. Each Sunday at 7:30 p.m. (E.S.T.), five kids each younger than sixteen received a $100 bond for answering questions asked by M.C. ("Master of Ceremonies") Joe Kelly. Several hundred kids appeared on the program over the years it was on the air, but some of them are still remembered by people who heard the show. Cynthia Cline, Sparky Fischman, and Gunther Hollander are names that still bring a nostalgic smile to the faces of *Quiz Kids'* fans. There was even a *Quiz Kids Game* which was sold in five-and-tens everywhere.

Other Panel Shows

Adult experts appeared on *Information Please.* This highly regarded program began on May 17, 1938 and ran until 1952. If a person mailed in a question the experts couldn't answer, he received a set of the *Encyclopedia Brittanica.* The experts included Oscar Levant (a pianist), John Kieran (a radio personality), and Franklin P. Adams (a columnist). A junior version was *Juvenile Jury* with five kids answering queries. *Life Begins at 80* did the same with oldsters. Another kid's show was *Uncle Jim's Question Bee.* What these shows had in common was they all used panels of experts—not amateurs.

Quiz Games

Amateur contestants appeared on many unusual quiz games. *Quiz of Two Cities* pitted contestants from one city against those from another by simultaneously broadcasting from two cities (a novelty during those days). *Rate Your Mate,* with M.C. Joey Adams had couples trying to guess whether their partners could answer a question correctly. As for sports quizzes, there was *Quizzer Baseball* with a pitcher and an umpire. *The Ask-It Basket, The Bob Hawk Show* (on which contestants won alphabet letters to form "LEMAC—Camel spelled backwards"), *Break the Bank* (with stunts like awards for the fattest and thinnest people), *Correction Please, Detect and Collect,* and *Double or Nothing* were shows that had a large following.

No doubt the most famous quiz show was *Dr. I. Q.* Lew Valentine was the M.C. (Dr. I. Q., the Mental Banker). Announcers would call out, "I have a lady in the balcony, Dr." Dr. I. Q. would then ask a question and if the lady answered correctly, she would receive a prize of ten silver dollars; if she lost, she got "a box of *Snickers* or *Milky Ways* and two tickets to next week's production." A highlight was a question about a famous person ("The Biographical Sketch") and there always was a memorable tongue-twister. There also was a kid's version of the show called *Dr. I. Q., Jr. Game Parade* was also a children's show; *Professor Quiz* was similar.

Musical Quizzes

In the late thirties and early forties, musical quiz shows were common. *Melody Puzzles, Sing It Again, Singo* and *What's the Name of*

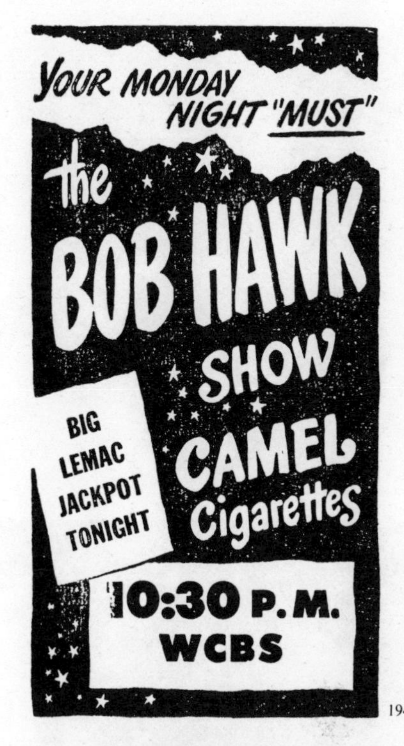

1948

TONIGHT
DON'T MISS
QUIZ KIDS
Right After You Listen to Jack Benny —Why not *Switch Your Dial* and Listen to the QUIZ KIDS?
WJZ—7:30 P. M.

1943

That Song? were all well known. Another musical quiz was *Grand Slam* which worked like a bridge game—the winner received a $100 savings bond.

Pot O'Gold with Ben Grauer and Horace Heidt and His Musical Knights asked listeners questions on the phone. It was this program that introduced the phrase "Stop the Music."

The musical quiz *Stop the Music* with Bert Parks as the M.C. offered a giant jackpot of $5,000 to anyone who guessed the mystery tune. The program was conducted over the phone. *Strike It Rich* also featured bigger prizes but most important was the *Heartline* which offered bonus prizes by listeners to those who were the most unfortunate or had the most heartrending hard-luck story.

Beat the Band was a musical quiz with a popular entertainer named The Incomparable Hildegard. The purpose was to keep the band from identifying a song by clues sent in from the audience.

A wacky quiz show was *Kay Kyser's Kollege of Musical Knowledge* with the musician asking contestants questions calling for answers of right or wrong instead of true and false. The result would be total confusion. If the answer was false and the contestant was correct, Kay Kyser shouted, "That's wrong, you're right."

More Quizzes

Hit the Jackpot with Bill Cullen was a popular show created by Bill Todman (who later produced other quiz shows on TV). Bigger money came with the *Sixty-Four Dollar Question* (earlier called *Take It or Leave It)*—each time the money doubled: $1, $2, $4, $8, $16. There were also *Speak Up, America, Spin to Win, Stop and Go.*

There was also some stress on educational quizzes like *The National Spelling Bee* with Paul Wing as the Spelling Master. Similar in format was *Noah Webster Says* which paid listeners two dollars for a list of five tough words to be used on the program, *Quick as a Flash, Scramby* (contestants unscrambled words), and *True or False.*

Truth or Consequences was once a fantastically popular program. M.C. Ralph Edwards asked contestants ridiculous questions which they never seemed able to answer. Then they had to pay the consequences, which were absurd stunts like bringing a streetcar conductor back to the show, frying an egg in the window of a big restaurant, or standing on their heads on Sunset Boulevard. Some of the contests which appeared in the forties were able to capture nationwide interest such as guessing the identity of Mrs. Hush who turned out to be a silent movie actress, Clara Bow, and identifying The Walking Man who was found to be Jack Benny. Prizes included a brand new Chevrolet. Most children waited eagerly with their parents for the Saturday night show with Ralph Edwards chortling "Aren't we devils?"

Twenty Questions also garnered a lot of listeners. This was a panel show whose panelists had to guess a subject within twenty questions after being told that it was either animal, mineral, or vegetable.

What Makes You Tick? was a psychology-oriented quiz show that enjoyed some popularity. The contestant rated himself and two psychologists rated him after asking some questions. Then there was *What Would You Have Done?* and *What's My Name?* and *Which Is Which?*

40 Kay Kyser ran a musical quiz show called *Kay Kyser's Kollege of Musical Knowledge.*

41 Professor Quiz! — (C. P. Corp.)

42 Ben Grauer

Ready for more? *Whiz Quiz, Who Said That?, Winner Take All, Yankee Doodle Quiz,* and *Youth vs. Age.* WHEW!

Groucho Marx did *You Bet Your Life,* a program later repeated on TV. Groucho was noted for his adlibs; however, they weren't really off the cuff—he interviewed contestants before the program—then made up jokes. The show is best remembered for the "Secret Word" (usually a common word like "bus" uttered by a contestant would win a prize.) As a quiz show, it wasn't much, but it was funny.

Reasons For Popularity

Quiz shows were the American Dream in miniature—the rugged individualist, pitted against the establishment, who won or lost because of his own knowledge or abilities. For example, if he knew a lot about a particular subject (such as geography), he might succeed; if he had nerve to keep going on *Double or Nothing,* or he were ingenious and brave enough to do stunts on *Truth or Consequences,* he could achieve modest fortune and limited fame.

The quiz shows were relatively cheap to produce but varied widely as successful entertainment. Producers had to keep coming up with unusual gimmicks to attract attention. Fred Allen at one point offered $5,000 to anyone who was phoned by *Stop the Music* and was listening to his show at the same time. Apparently, he never had to pay off though he had many people who pretended that they missed out by listening to him.

No one ever investigated the honesty of radio quiz shows. That was to come later during the television quiz scandals when some big money winners admitted they were given the answers before the show began. The public was shocked.

But the radio quiz show contestant who received silver dollars may have been just as dishonest. At least one person has claimed to have seen members of the audience paying ushers in order to be chosen as contestants.

At any rate, many of the radio quiz shows were transferred to television. Many of the stunts in radio shows like *Truth and Consequences* had to be verbal or at least capable of being easily described. TV permitted complex stunts, usually in the studio where the audience could see them, in contrast to radio which permitted action out of the studio with later descriptions. Telephone quizzes never became quite as popular on television as they had been on radio.

How Quiz Shows Worked

No doubt a better name for these programs is the present one: game shows, because they are more of a game than a quiz. Much emphasis was placed on the unique individuality of the contestant. Listeners particularly enjoyed a contestant who had just been married, had an unusual hobby like collecting shoelaces, or had a curious way of speaking or acting. Since (in the early days) these people were chosen from the audience, quiz shows excelled simply because, by luck, better contestants were chosen.

The procedure worked like this. A half-hour or so before the show began, the audience filed in and were given cards to fill out which asked questions about where they lived and what interests or hobbies they had. A staff announcer (not the M.C.) came on and read names from the cards or selected people at random from the audience. This "warm-up," as it was called, gave the audience a

43

44

43 *Information Please* began on May 17, 1938, with Franklin P. Adams as moderator. The show opened with a rooster crowing and the announcement, "Wake up America; it's time to stump the experts!"

44 Art Linkletter was the MC of many radio shows. Here he appears on *House Party,* which ran Monday through Friday from 3:30 to 3:55 p.m. (EST) on CBS. — (CBS photo)

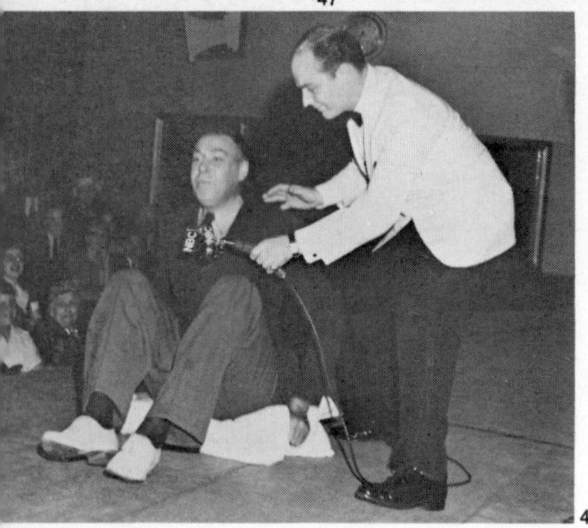

45 Ruthie Duskin, the eight-year-old literature expert of the *Quiz Kids.*

46 Ralph Edwards as host for 1941 *Truth or Consequences.* — (NBC photo)

47 Joe Kelly and a Quiz Kid, 1940. — (NBC photo)

48 The consequences in *Truth or Consequences.* — (NBC photo)

chance to relax and to get to know the contestants and to adjust to the show's atmosphere.

Because selection of contestants from the studio audience was risky (the audience might be full of dull people), some programs began soliciting letters from would-be contestants. These letters were carefully read, possible contestants interviewed in the network offices and selected contestants asked to return. Thus, a balance program of contestants could be selected, as, for example, a very witty person and an average one, equal numbers of men and women, one old person and one young one and so forth. After all, the networks reasoned, they were selling entertainment and there was no reason they shouldn't take every precaution to make sure that they had covered all the bases. (It was this kind of reasoning, applied to giving answers to insure the popular person winning, that unfortunately led to the TV quiz show scandals.)

Although the prize given on radio quiz shows was usually money, someone discovered that merchandise often was much more attractive. That is, a brand new automobile seemed much more glamorous and desirable than the $1,500 it cost in those days. Further, it was usually possible to promote those prizes, that is, to get them free in exchange for their advertising value; a manufacturer would supply the show with a car and it would cost him less than a commercial. In addition to getting the make of car mentioned frequently, the prize had a large word-of-mouth audience; people thought of winning a new Chevrolet (not just a new car).

The key to a good quiz show was the pace. If the M.C. could maintain a fast moving show, the program attracted a large listening audience. Someone thought of getting the audience involved in speeding up the show and making it seem breathless. One way of doing this was with lots of applause. An "Applause" sign was installed which blinked on and off wildly at every possible instance. Some members of the production crew were told to applaud madly and wave at the audience and thus encourage them to join in. Someone invented an applause machine which created the sound of hands clapping by means of rotating a drum with wooden strips attached to it. A canned laughter machine came next.

The All-Important M.C.

Probably the most difficult job in radio was to be the M.C. of a radio quiz show. The M.C. had to be very friendly and enthusiastic. Further, he had to be prepared for almost anything to happen on mike. A contestant might faint, utter profanity, or suddenly become either mute or too talkative. Further, the M.C. had to keep the sympathy of the audience when the contestant lost, for the audience usually felt that the contestant should win and they often felt cheated along with him if he did not.

The Questions

The content of the questions bears mention too. The audience liked to be slightly mystified and pleased when someone could get the answer. They seemed extremely good judges of what was a fair question or not. If a question was too difficult or the answer was very close they were easily upset. It would be disastrous for the program if they booed. Thus, there was a qualified, or, at least, certified expert, usually a college professor, who determined whether the answer was right or not.

Quiz Shows: That's All?

As TV quiz shows increased, radio's decreased. In time all that was left were announcers dialing names out of a phone book to see if they were listening and could name some tune for the jackpot of the week. Old radio quiz fans shook their heads sadly and turned the TV dial.

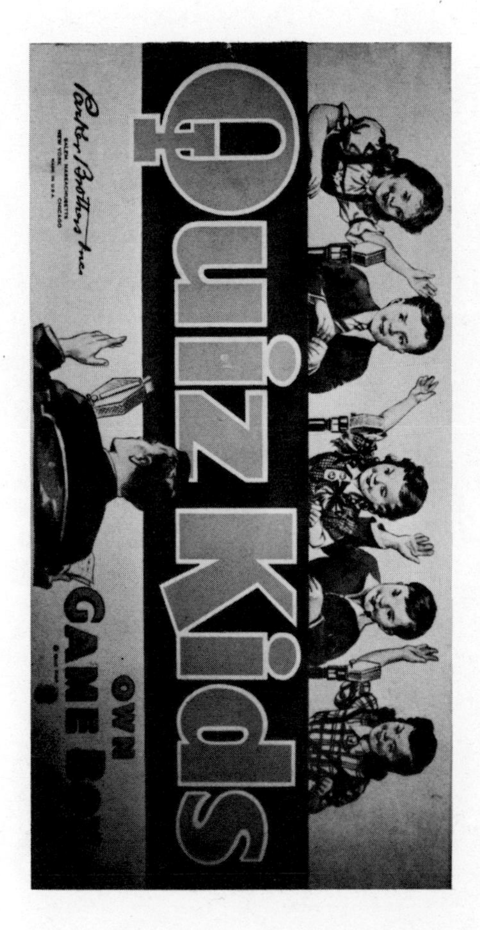

The Quiz Kids

This panel show consisted of very bright youngsters who had to retire from the broadcast when they reached 16.

M.C.—Joe Kelly
Announcer—Ken Carpenter
Quiz Kids—Richard Williams (11),
Jack Lucal (14),
Joan Bishop (14),
Claude Brenner (12),
Gerard Darrow (8),
and on this show, the comedian, Jack Benny.

CARPENTER.—Here they are—The Quiz Kids!—presented by the makers of Alka-Seltzer. We're on the air with the School Kids' Questionnaire!

MUSIC.—*(Organ). Theme.*

CARPENTER.—The Quiz Kids! Five bright, lovable youngsters, ready for another difficult examination in the Alka-Seltzer Schoolroom of the Air. The examination tonight will be conducted in exactly the same manner as all our regular Wednesday night Quiz Kid programs, and, as usual, none of the children have seen or heard any of the questions in advance.

BENNY.—I'll say we haven't. Let's get going.

CARPENTER.—All questions were sent in by you listeners and were selected by Sidney L. James, of the editorial staffs of *Time* and *Life* magazines.

BENNY.—I don't care who sent them in. Let's get going. I can answer them, you know.

CARPENTER.—A new Zenith portable radio will be awarded the sender of each question used on this program tonight. And now our chief quizzer himself . . . Joe Kelly!

SOUND.—*Applause.*

KELLY.—Thank you, Ken Carpenter, and good evening, ladies and gentlemen. Well, we'll proceed directly to the roll call. Richard . . .

RICHARD.—I'm Richard Williams. I'm eleven years old, and I'm in the sixth grade at Harrison School, East Chicago, Indiana.

KELLY.—Jack Lucal . . .

JACK.—I'm Jack Lucal. I'm fourteen years old, and I'm a freshman at the Oak Park and River Forest Township High School.

KELLY.—Joan . . .

1941

JOAN.—I'm Joan Bishop. I'm fourteen years old, and I go to the Chicago School for Adults.

KELLY.—Claude . . .

CLAUDE.—I'm Claude Brenner. I'm twelve years old, and I am a sophomore at Senn High School in Chicago.

KELLY.—Gerard . . .

GERARD.—I'm Gerard Darrow. I'm eight years, and I go to the Bradwell School on Burnham.

KELLY.—And, Jackie . . .

BENNY.—I am Jackie Benny. I'm six years old . . . ah—I didn't have a chance to go to school at all.

SOUND.—*Laughter.*

BENNY.—I was just a poor boy, and I used to stand on the corner selling papers *(laughter),* barefooted in the winter, and I used to say, "Extra . . . extra . . . paper here . . . get your paper . . ."

KELLY.—Quiet, please.

BENNY.—Hmm . . . Fine chance I'm going to have here, I can see that . . . You know . . .

KELLY.—Now, Jackie, please.

BENNY.—I know just as much as the kids, you know. You just ask the questions, that's all.

KELLY.—Jackie, please. And incidentally, where are your curls?

BENNY.—What?

KELLY.—Where are your curls?

BENNY.—On my lap. They got hot.

SOUND.—*Laughter.*

KELLY.—Well, while we're getting ready for our first question, just a word or two from Ken Carpenter.

(Commercial)

BENNY.—You said it.

KELLY.—Quiet, please. We'll now start with the questions. All right, Quiz Kids . . .
R. S. Hart, of Seattle, Washington, says he was in the desert and after making an analysis of the only water available showed that it was 100 per cent aqua fontis. Would you drink such water? Joan . . .

JOAN.—Yes I would.

KELLY.—Well, can you give us anything further?

JOAN.—Well, aqua fontis is fountain water.—

KELLY.—That's right . . . Well . . . it's really spring water, Joan.

JOAN.—Oh.

BENNY.—Yeah, it's *spring* water, Joan.

SOUND.—*Laughter.*

BENNY.—That's right, Mr. Kelly, it's spring water.

KELLY.—Yes, I know. It says so on my card.

BENNY.—Oh, that's where I saw it before.

SOUND.—*Laughter.*

KELLY.—All right, the next question.
Pete McDonald, of Veronia, Oregon, a schoolboy, who says that he never enjoyed anything in school but recess until he began listening to The Quiz Kids, sends in this one. Incidentally, he adds that his grades are improving. Here it is—if you had something that contained a prothorax, a mesothorax . . .

BENNY.—A what? . . . a mess o' what? . . . What did you say . . . a mess o' what . . .

KELLY.—A mesothorax.

BENNY.—Oh, a mesothorax.

KELLY.—And a metathorax, what would you have? Gerard . . .

BENNY.—Gerard, you answer. You had your hand up first.

GERARD.—Now, Mr. Benny, don't butt in, *please.*

BENNY.—Well, that's something . . . I can see I'm certainly going to have a fine chance here today.

KELLY.—All right, Gerard.

GERARD.—The metathorax, the mesothorax, and prothorax are all part of the thorax, which is part of an insect on the . . . the thorax is the part between the abdomen and the head on an insect.

KELLY.—Well, good for you, Gerard. That was marvelous.

SOUND.—*Applause.*

KELLY.—That's very good.

BENNY.—I used to know that when I went to school . . . You know, when you get older you forget those things. You can't remember everything . . .

KELLY.—Now, our next question.

BENNY.—I used to know algebra, too, when I went to school.

KELLY.—Quiet, please.

BENNY.—Oh . . .

KELLY.—Mrs. Burdett E. Truedson, of New York City, says you can prove you have a good background by naming at least three persons whose names will live forever because their names have been used to identify their chief contributions to humanity. For example, the name of Roentgen is perpetuated in the word "roentgenology" . . . Claude . . .

1946

1936

CLAUDE.—Nobel . . . he was a Swiss scientist who discovered dynamite and . . . he . . . people . . . he gives out prizes to people who do something great for the world.

KELLY.—That's fine, Claude. Let's see what Joan has to offer.

JOAN.—Well, there's Calvinism . . . That's the doctrine as to the downfall of man . . . and Darwinism . . . ah . . . the theory of anthropology.

KELLY.—Very good, Joan. Jack Lucal.

JACK.—There's Alessandro Volta . . . his name is perpetuated in the volt, by which we measure electricity. And James Watt . . . they use his name, too, for the watt.

SOUND.—*Applause.*

KELLY.—Nice going, Jack Lucal. Let's see . . . Richard.

RICHARD.—Well, Martin Luther in the word "Lutheran," which is a church, and Dr. Roentgen who discovered the Roentgen rays.

SOUND.—*Applause.*

KELLY.—That's very good, Richard. Jackie has his hand up. What . . .

BENNY.—Well, there's a fellow named Max . . . he had something to do with the Maxwell . . . the Maxwell . . .

SOUND.—*Laughter.*

KELLY.—Now, wait a minute, Jackie . . . there's no connection there.

BENNY.—There is, too, a fellow named Max . . . sold me my car . . .

SOUND.—*Laughter.*

BENNY.—Max . . . Maxwell . . . his name was Max Miller . . . I've certainly got a fine chance on this program . . . I should have stayed on my own Jello show.

KELLY.—Well, it's beside the point, but we'll accept it as half right. Claude . . .

BENNY.—Oh, well. It's about time.

CLAUDE.—Also there's Andre Marie Ampere . . . He had something to do with electricity, and his name lives in the ampere.

BENNY.—Oh, the ampere, the ampere . . .

SOUND.—*Laughter.*

KELLY.—That's right, Claude. Jack Lucal.

JACK.—Well, Cadillac and La Salle were French explorers, and their names are names of automobiles.

SOUND.—*Laughter.*

KELLY.—Very good, Jack Lucal . . . I guess that will hold Jackie for a while. All right, our next question.

BENNY.—If everybody is going to get laughs on this program. I'm going home.

SOUND.—*Laughter.*

KELLY.—Gerard . . .

GERARD.—Well, there is also De Soto, whose name is a car . . .

SOUND.—*Laughter.*

GERARD.—He was the Spanish explorer that found the Mississippi.

KELLY.—That's right, Gerard. I am glad you brought that up.

BENNY.—What about Johnny Chev that made the Chevrolet . . .

SOUND.—*Laughter.*

BENNY.—For heaven's sake, if you're going into that kind of stuff . . . you know . . . Johnny Chev . . . What about Harry Stu . . .

SOUND.—*Laughter.*

BENNY.—With that stuff I can answer a million of them . . . you know . . . just ask some questions, that's all.

KELLY.—We'll all withdraw from the garage right now . . . and get into our next question.

SOUND.—*Laughter.*

KELLY.—Miss Margaret Faith, of Camden, New Jersey, poses this mountain climbing and mathematics problem.

BENNY.—*(Mumbling)*

KELLY.—A mountain climber was making his way along a mountain-side ledge . . .

BENNY.—Wait. Pardon me, who was it asked the question, please?

KELLY.—Miss Margaret Faith, of Camden, New Jersey . . .

BENNY.—Oh . . . Camden . . . yes . . . yes, I see.

KELLY.—Let's see, where are we? Oh, yeah.

SOUND.—*Laughter.*

KELLY.—A mountain climber was making his way along a mountain-side ledge at an altitude of 6,440 feet. While edging his way, he accidentally kicked a rock which went flying toward the bottom of the mountain at some animals who had to scurry for shelter. Ignoring the friction of the air, how long did the animals have to reach safety before the rock hit? Now, you have to do this in your head, kids. No pencil and paper.

BENNY.—What is the last question again, please? How many . . . how long did it take what?

KELLY.—Well, that's the question . . . how long did it take?

BENNY.—One minute and 43 seconds.

KELLY.—That's wrong . . . Richard.

BENNY.—I've certainly got a fine chance here.

KELLY.—Richard.

RICHARD.—Twenty seconds.

KELLY.—Twenty seconds is correct.

SOUND.—*(Applause)*

9:00—WJZ

CHARLES LAUGHTON

GUEST STAR

1946

SPECIAL !
AMATEUR CLASSES
AT THE
Eastern Radio Institute
Start Saturday, October 8th
10 A. M. to 12 Noon

Consists — Code, Traffic, Radio Theory, Design and operation of apparatus. All members entitled to Special Discount at F. D. PITTS CO., 12 Park Square, Boston.

New members may join any Saturday.

Tuition very reasonable! For further particulars write or call at

Eastern Radio Institute
899 Boylston Street
Boston, Mass.
Tel. Back Bay 5964
F. D. PITTS, Director

1921

BENNY.—Well!

SOUND.—*Bell.*

BENNY.—No wonder. He *squared* the root. I tripled it.

SOUND.—*Laughter.*

KELLY.—Well, nice going, kids, and though I don't think you need it, you can rest a while. It's recess time.

(Commercial)

MUSIC.—*(Organ). Theme . . . final eight bars to ending.*

CARPENTER.—Ladies and gentlemen, you are listening to The Quiz Kids, presented every Wednesday night at this time by the makers of Alka-Seltzer. Now, just a word about the questions. You can win a new Zenith portable radio with patented, built-in wave magnet if you send us a question which our question editor finds suitable for use on the air. Yes, Alka-Seltzer awards a famous Zenith portable radio for each question used on this program. Just mail your questions by post card or letter to Quiz Kids, National Broadcasting Company, Chicago. We reserve the right to reword questions, and if like questions are submitted, the first received will be used. All questions become the property of Quiz Kids. So send in your question, and win a radio.

BENNY.—You better see that I get the $100 bond, too . . . that's all I'm worried about.

CARPENTER.—All right, Joe, are you ready with the scores at the halfway point?

KELLY.—Yes, Ken, but in deference to our guest contestant, I hesitate to read them. I think I'll just let them go until after the second question session. Maybe a miracle will happen. By the way, Richard, that last question we had before the bell, can you tell us how you worked that out?

RICHARD.—Well, Mr. Kelly . . . any body falling through space, disregarding the friction of the air, accelerates at the rate of 32.2 miles . . . uh . . . feet per second, and so the rule is, the distance equals the time in seconds squared times half the acceleration per second, and in this case it was 6,440 feet equals 16.1 times the time squared, so I divided 6,440 by 16.1 and got 400, which is the square of the time in seconds, and I extracted the square root, and that gave me 20, and so the answer is 20 seconds.

KELLY.—Good for you, Richard, my boy.

SOUND.—*Applause.*

KELLY.—Now let's see . . .

BENNY.—Where I made *my* mistake there, see . . . I took the least common multiple . . .

SOUND.—*Laughter.*

BENNY.—*That's* where I got wrong. That's where I got the minute 43 seconds.

SOUND.—*Laughter.*

KELLY.—You sort of squared it there.

BENNY.—That's what I said.

KELLY.—Well, let's get along here now. Here's a question from Mrs. Daniel Stormont, of Evanston, Illinois.

BENNY.—5,280 feet is 1 mile.

KELLY.—What? What did you say?

BENNY.—I said 5,280 feet is 1 mile.

KELLY.—Well, nobody asked that one.

BENNY.—Well, if they *do*, I'm ready, Watch out.

SOUND.—*Laughter.*

KELLY.—All right, we'll continue.
If you told the election board you were a mugwump, would you be listed as a Republican, Democrat, Socialist, or Independent?

BENNY.—I wouldn't tell *anybody* I was a *mugwump.*

SOUND.—*Laughter.*

KELLY.—Well, Claude.

CLAUDE.—I'll take a guess. I'd say an Independent.

KELLY.—That's right, but how did you guess it?

CLAUDE.—I just guessed.

KELLY.—Oh, you just guessed, that's right. You see the political name of mugwump . . . well, let's see what Joan has to say.

JOAN.—Well, I rather thought it was Independent too, because there's a column in one of our Chicago papers called "Mugwump."

KELLY.—That's true . . .

JOAN.—On politics.

KELLY.—You see, the political name of "mugwump" started in 1884, when it was applied to the political supporters of James G. Blaine, who switched to Cleveland because of his civil service views. Blaine was Republican candidate for President . . . Jackie . . .

BENNY.—I know what a mugwamp is.

JOAN.—Oh, oh.

KELLY.—You do? All right.

BENNY.—A mugwump is a bird that sits on a fence with his mug on one end and his wump on the other.

SOUND.—*Laughter and applause.*

GERARD.—Mr. Benny, I'm afraid you're wrong.

1929

RADIO FACTS!

Religious Programs

No discussion of radio is complete without mentioning religious programs, which occupied most of the schedule on Sunday morning and afternoon with as large a following as any entertainment show. In addition to broadcasts by local ministers, priests, and rabbis, national prominence was attained by some, like Father Charles Coughlin and Harry Emerson Fosdick. Some long lasting programs were *The National Church of the Air, The Old-Fashioned Revival Hour, The Voice of Prophecy, The Lutheran Hour, Young People's Church of the Air,* and *The Hour of Faith.*

Religious dramas included *The Greatest Story Ever Told, The Guiding Light,* and *The Light of the World.*

KELLY.—Well, let's have a little more discipline, please. Getting back to the political situation, Jackie Benny, who was President of the United States in 1901?

BENNY.—Grover Cleveland.

KELLY.—That's wrong.

BENNY.—Well, I ought to know. I voted for him.

SOUND.—*Applause.*

BENNY.—It was Grover Cleveland.

KELLY.—You're wrong. It was William McKinley.

BENNY.—I just wish I had a history book, brother, that's all.

KELLY.—I've got one.

BENNY.—Well, give it to me, I've got a low chair here . . . It was Grover Cleveland, that's who it was.

KELLY.—Let's continue with the next question. Pauline Salzman, of Grand Rapids, Michigan . . .

BENNY.—It was Grover Cleveland. I know Grover Cleveland . . .

KELLY.—Quiet, please. I'd like to present this question. Pauline Salzman, of Grand Rapids, Michigan, found these ads in the paper. She would like you to tell her just what is advertised. Here is the first item: "For rent. Colonial estate near Charlottesville, Virginia. Designed by owner. Adjoining buildings make estate virtually a community. Write owner—T. J., Charlottesville, Virginia." Jackie Benny you're holding your hand up.

BENNY.—I'm waving at some friends in the audience. I can have friends in the audience, can't I? Hello, Mamie!

SOUND.—*Laughter.*

KELLY.—Well, let's complete this question. Richard.

RICHARD.—Monticello.

KELLY.—Monticello, the home of . . .

RICHARD.—Thomas Jefferson.

KELLY.—That's right. Good for you.

BENNY.—That's right.

SOUND.—*Applause.*

KELLY.—Here's the next item: "For sale . . ."

BENNY.—You can have friends in the audience you know . . . good heavens . . . otherwise, there's no use in being here . . .

KELLY.—Quiet, quiet, please. Here's the second part of this question . . . "For sale. Sacrifice. Ten million dollar marble home in Land of Veda. Stands on 313-foot-square marble terrace. Absolutely unique as architects' eyes poked out after construction completed" . . . Claude.

CLAUDE.—That's the Taj Mahal.

KELLY.—The Taj Mahal in India. Good for you.

BENNY.—It took 22,000 men 22 years to build it.

SOUND.—*Laughter.*

BENNY.—And I'm right about Grover Cleveland, too . . . I guess I know about Grover Cleveland . . . you know.

KELLY.—Well, we'll continue . . . Frank O. Estes, of Towson, Maryland, sends in this one. Last Christmas his wife went shopping to get her girl friends gifts. She bought Grace a green umbrella for $2.95, Ellen a blue scarf for $2.50, Jo Anne a brown leather pocketbook for $2.99, and Priscilla a yellow sport skirt for $3. What was the color of the scarf for Ellen? . . . Joan.

JOAN.—Blue.

KELLY.—Blue. That's right. Good for you.

BENNY.—1901 *was* Grover Cleveland.

SOUND.—*Laughter.*

BENNY.—I know, because I won a pair of cloth-topped shoes on the election . . . I remember that . . .

KELLY.—We'll forget about Grover Cleveland.

BENNY.—I won't forget about him.

SOUND.—*Laughter.*

KELLY.—Well, this next question here . . .

BENNY.—It burns me up. Come over here and you . . .

KELLY.—Quiet, please. Connie Haitomt . . .

BENNY.—Connie Haitomt . . . Connie Haitomt . . .

SOUND.—*Laughter.*

KELLY.—Now, Jackie, I'm reading a name . . .

BENNY.—All right, read the name.

KELLY.—All right, quiet. Connie Haitomt . . .

BENNY.—I'm not getting paid for this, you know . . .

SOUND.—*Laughter.*

BENNY.—I just came over . . . I'm just a guest . . . that's what burns me up . . . you know.

KELLY.—Connie Haitomt . . . Listen, Jackie, I'm beginning to think you're getting into what little hair I've got left.

BENNY.—I can always tell you where to get a toupee, you know.

SOUND.—*Laughter.*

KELLY.—Quiet. Connie Haitomt of Minneapolis, Minnesota, wants you to sing or hum these notes as I give them to you, and stop me as soon as you recognize the scales you are singing. All right, here is the first one: C, D E♭, F, G, A♭, B, C.

JOAN.—*(Sings)*

1949

SIR HUBERT
WILKINS
tells of his daring under-ice exploit tonight, 9:15 P. M.
"BAYUK STAG PARTY"
WJZ, WLW, KYW, KWK, WBAL, WGAR, WREN

1931

PHILCO

BABY GRAND
RADIO SET

$49.95

Complete

7 POWERFUL
TUBES

1931

DYNAMIC SPEAKER

SUPERHETERODYNE
CIRCUIT

More than any other set at the price has ever offered.

KELLY.—All right, Joan, do you recognize the scale?

JOAN.—That's the harmonic minor.

KELLY.—That's very good. Two: C, C#, D, D#, E, F, . . .

BENNY.—If I had my violin here I'd have gotten it.

KELLY.—I'll tell you what we're going to do. We've got some other hands up. I'm going to give this one . . . to Claude.

CLAUDE.—That's the chromatic.

KELLY.—Chromatic is correct. Good for you Claude. And here is the last one: C, D, E♭, F, G, A, B, C, B♭ . . . Richard.

RICHARD.—That's the melodic.

KELLY.—Melodic is good. Good for you kids.

SOUND.—*Applause.*

BENNY.—It's one of the *silliest* questions I've ever heard.

SOUND.—*Laughter.*

BENNY.—*(Mumbling)* Once in a while . . .

KELLY.—Now, Quiz Kids, you'll need mythology . . .

BENNY.—I raise my hand all the time, and nobody even calls on me . . .

SOUND.—*Laughter.*

KELLY.—You'll need mythology as well as ornithology to answer this one. Ethel Baker, of St. Louis, Missouri, wants to know why peacock feathers are spotted.

BENNY.—Is that Paul Baker's sister?

SOUND.—*Laughter.*

KELLY.—What?

BENNY.—It isn't Paul Baker's sister, is it?

KELLY.—When you want to talk, Jackie, will you please hold up your hand?

BENNY.—. . . because I *know* a Ethel Baker, you know . . .

KELLY.—Well, remember to hold up your hand when you want to say something.

BENNY.—All right, I'll hold up my hand. For heaven's sake what does he think he is . . . the boss or something?

KELLY.—Quiet, please.

BENNY.—It's the last time I'll come on this show . . .

KELLY.—*You're telling us!*

SOUND.—*Laughter.*

KELLY.—Now, let's see, where am I? . . . Ethel Baker, of St. Louis, Missouri, wants to know why peacock feathers are spotted. Gerard.

GERARD.—The peacock has eyes in its tail feathers because . . . ah . . . it's a mere myth . . . you see when . . . a long time ago when Jupiter married Juno, after a few years he became jealous of her and turned her into a calf, and he sent Argus to watch, but Juno turned herself right back into her regular form, and Argus was the one that had a hundred eyes in . . . his head, and Juno killed Argus and put the eyes in the peacock's tail.

KELLY.—Well, thank you very much, Gerard.

SOUND.—*Applause.*

KELLY.—That was a very fine description . . . Jackie, I see you've got your hand up.

BENNY.—I'm wiping my forehead. It's hot in here. You can't even raise your hand. Most ridiculous questions I ever heard.

KELLY.—We'll continue. James Wilson, Jr., of Toledo, Ohio, wants you to compose a second line to his one-line verse. Here it is—"Fred Allen has a funny show . . ."

BENNY.—*I'm going home.*

SOUND.—*Laughter.*

KELLY.—You keep your seat . . . All right . . . "Fred Allen has a funny show" . . . Let's hear a second line to that . . . Joan.

JOAN.—Fred Allen has a funny show, But there's not a thing he doesn't know . . . Hum!

KELLY.—Very good, Joan. All right, let's have another one.

BENNY.—What's funny about that?

KELLY.—Gerard.

GERARD.—When Mr. Benny hears that he'll surely blow.

SOUND.—*Laughter.*

KELLY.—All right, Jackie, what do you have to offer?

BENNY.—Fred Allen has a funny show.
How he does it, I don't know.
His jokes are old, his gags ain't funny.
He ought to be paid in Confederate money.
The end.

SOUND.—*Laughter.*

KELLY.—Now, then, here is really one for you, Jackie Benny.

BENNY.—Fine, my father is listening in.

KELLY.—All right, Jackie, how many strings on a violin?

BENNY.—Five, I mean four.

KELLY.—Very good . . . How do you spell rosin?

BENNY.—R-O-S-O-N.

KELLY.—That's wrong. It's R-O-S-I-N.

BENNY.—I can't understand it. I've been using it for years.

1942

SOUND.—*Laughter. Bell.*

KELLY.—Well, there's the bell, kids. I'll have your scores in just a moment.

MUSIC.—*(Organ). Theme . . . final eight bars to ending.*

KELLY.—Well, kids, as a group you missed only one question tonight, and the individual winners are Richard—first, Joan—second, and Claude—third.

I congratulate all you Quiz Kids and take pleasure in presenting to each of you, in behalf of the makers of Alka-Seltzer, a $100 denomination United States savings bond. Jackie Benny, I don't have one for you. You see, these bonds are to help the children pay for their future education, and we didn't think you'd spend your money in going to college. But here's a Zenith portable radio. Maybe you can learn something listening to the Quiz Kids every Wednesday night.

BENNY.—Well, at least I can hock the radio.

KELLY.—Well, we'll be back in Chicago next week, and we'll resume competition with only the three highest scorers remaining for the succeeding examination. The three winners on our last competitive program were Claude, Richard, and Jack. Completing the board will be Gerard and Joan, the same children on the program tonight. Meanwhile, this is Joe Kelly dismissing the Quiz Kids class until next Wednesday at the same time. Good night, kids!

Keep cookin' with Crisco.
It's all vegetable.
It's digestible.

KIDS.—Good night, Mr. Kelly.

BENNY.—Come on, ask some more questions. Let's get going here! Come on . . .

MUSIC.—*(Organ). Theme.*

CARPENTER.—Listen again next Wednesday night to The Quiz Kids. The makers of Alka-Seltzer present three programs each week . . . all of them on NBC networks. On Friday night, Alec Templeton Time; on Saturday night, the famous Alka-Seltzer National Barn Dance; and next Wednesday night again, The Quiz Kids. For interesting variety and entertainment, listen to the Alka-Seltzer shows. Ken Carpenter speaking.

ANNOUNCER.—This is the National Broadcasting Company.

1942

MUSIC
MUSIC
MUSIC

Chapter 7

49 Morton Downey, popular tenor. (CBS photo)

50 Kate Smith ("The Songbird of the South") began her show with "Hello, everybody." — (CBS photo)

51 "Der Bingle" — Bing Crosby. — (Bing Crosby Enterprises, Inc., 1949)

Do Re Mi . . . Know anybody who doesn't like music? Of course not. In fact, music was once so popular on the airwaves that musicians feared radio would put them out of work. After all, they reasoned, who would buy a record or come to a concert when they could hear free music on the radio? The Musicians' Union bitterly fought with station management in the thirties and forties, finally winning agreements which assured them employment.

Earliest Programs

Although Eugenia H. Farrar was the first known vocalist to sing over the air (December 16, 1907 from the Brooklyn Navy Yard), the earliest music program was *The Lady Esther Serenade* which began on September 27, 1931. Harry Horlick and the A&P Gypsies were the stars of an early musical program with Frank Parker, a tenor who later sang with *Arthur Godfrey and His Friends.* Milton Cross, announcer on the Saturday afternoon *Texaco Opera* broadcasts, was also on the show.

In the early days, most singers worked free. One station did pay Harry Richman $1,500 to sing three songs, but that was a rare exception. Even the famous Kate Smith got no pay because she thought that "it helped her make money on personal appearances." Kate's bubbling enthusiasm (she was the Mama Cass of her day) was infectious. She was patriotic (*God Bless America* was a great success for her) and a great saleswoman (her effort at selling war bonds over the air is legendary). Above all, she had a distinctive, fine voice.

Those musicians lucky enough to have sponsors used names which identified them, like the A&P Gypsies, the Ipana Troubadors, and the Interwoven Boys (Billy Jones and Ernie Hare). Exceptions were Paul Whiteman and His Rhythm Boys and cigar smoking Ben Bernie (famous for shouting, "Yowsah!"), who conducted "The Lads."

The Crooners

Harry Lillis "Bing" Crosby left Paul Whiteman's Rhythm Boys when CBS offered him $1,500 a week for five fifteen-minute broadcasts. To compete, NBC hired Russ Colombo, another "crooner" (a new style of singing that was all the rage.)

The radio microphone made crooning possible—a singer stood close to it and used soft tones which couldn't be heard unless they were amplified. Although Al Jolson and others had made fortunes with popular music, their singing style was traditional. Curiously, the crooning of Crosby, Colombo, Skinnay Ennis and others made people think that they were undermining the morals of youth.

Country Music

In the 1920s, America was still a rural country and this was one of the reasons for the increase in popularity of country and western music. WSM in Nashville, Tennessee, began the *Grand Ole Opry* (a program which exists today). The singers from the hills were paid little or nothing to sing on the station but believed that they soon made it up in record sales and personal appearances. Many of them were not full-time professional entertainers but owned ranches and farms to make ends meet. Some, like the legendary Uncle Dave Macon, a banjoist, began his show business career late in life but rapidly became well known to a specialized audience.

Country music was heard throughout the southwest from the powerful stations owned by "Doc" Brinkley, who was forced by the FCC to sell his patent medicine selling station KFKB in Milford, Kansas, which (incredibly) was the most popular radio station in the United States in 1930. He moved into Mexico and began broadcasting from unregulated stations back across the border. Through these stations (XERF was one), the music of the Carter family and other country musicians reached rapid and widespread popularity.

This music, with its simple lyrics and foot-stomping melodies, reached a select audience. There were musicians whose names were household words in the South and West, but who were unknown to those who lived above the Mason-Dixon Line.

The East had no such regional music, and its stations rapidly became part of the chains or early networks. Thus, their audiences were not the captives of a strong independent and specialized station like WSM (Nashville) or XERF (Del Rio, Texas). Eastern stations searched for a national music like the A&P Gypsies or Jessica Dragonette that would suit the tastes of most of its listeners.

It's a Hit

Thanks to radio, a song would become a hit overnight. Music lovers were anxious to keep up with the latest hits and *Lucky Strike's Hit Parade* capitalized on this with a weekly program of hits. This list was based on figures from record sales and sheet music sales as well as the number of times that a song was played on the juke box and the radio.

The show created suspense around the top three tunes. According to Giselle MacKenzie, a singer who appeared on the show, frequently the cast did not know what the Number One Song was until seconds before they were to sing it.

When a song reached the Number One Spot, the program often helped keep it there; listeners raced out to buy sheet music and records. The *Hit Parade* had a wide and varied audience; a 1942 study revealed 54% of the nation's farmers and 70% of the country's urban dwellers listened each Saturday night to the Ten Top Hits.

Other shows used the gimmick of having experts guess what tune might make it as a hit. The majority of the songs dealt with love (usually unrequited) and almost all were in 4/4 time. The war popularized patriotic songs like *Coming In On a Wing And a Prayer* and *Praise the Lord and Pass the Ammunition.*

Legal Problems

"Canned" or recorded music was a problem. The Supreme Court ruled that playing music over the air was a copyright infringement; that is, if royalties were not paid to the publisher, the station was liable to a payment of $250 for each violation.

James C. Petrillo, president of the Chicago Federation of Musicians, started a fight against WCRW for playing records over the station. The owner, Clinton R. White, invented a device called the Vibrophone which broadcast records without putting a mike in front of the player, which gave the listener the impression he was listening to a live orchestra. The Federal Radio Commission did not approve of the idea and warned him that this was a violation of copyright.

Worse, some unscrupulous stations continued to use records and pretended that the musicians were there in person. According to one

52

53

54

52 "Frankie" (In the forties, bobbysoxers swooned over him.)

53 Sarah Vaughan, Vocalist.

54 One of the greatest showmen of all time was Al Jolson, who met with great success as a singer on radio.

report, announcers were instructed to continue talking as the record began to cover up any needle scratches. Just before the depression, the FRC put an end to this by requiring announcers to state that they were using recordings.

Make-Believe Ballroom

On February 3, 1935, Martin Block, a WNEW announcer, thought about filling the air between bulletins from the famous Lindbergh kidnap trial; he decided to play some Clyde McCoy records. Because of this, most books refer to him as the first disc jockey. (However, in 1910, an engineer, Dr. Frank Conrad, played records by request on his experimental station.) Martin Block got his own sponsor, Retardo, a reducing pill and he was an immediate success with the new program he called "Make Believe Ballroom." As a salesman he became a legend; he sold 300 refrigerators for Bamberger's Department Store during a 1938 blizzard. Later during World War II he asked for donations of pianos to the USO and got 1,500.

Music occupied more than one-half of all air time, more than any other type of programming. It was the disc jockeys more than the live musicians who made music important on radio. True, the listeners tuned in to hear Dinah Shore, Bing Crosby, and Frank Sinatra. But they listened to them in person only because they had heard them on records. And they heard the records first on radio. Disc jockeys thought up unique ways to involve their listeners in their programs—from taking phoned-in requests for songs ("Now for all the gals at Olney High . . .") to broadcasting from unusual spots like atop a flagpole. Pop music accounted for two-thirds of music programming.

Rock 'n Roll

When rock and roll started in the mid-fifties, with the recordings of Bill Haley and the Comets, the disc jockey began to emerge as a very important figure. Alan Freed had started in Akron, moved on to Chicago, and then on to New York where he literally created rock 'n' roll (at that time named rhythm and blues). He had a great knack for promotion and many singers made a success of their recordings after a playing on his show. He literally made hits.

The King

In 1956 disc jockeys were talking about a young kid from the south who wore pink shirts and pink slacks, had long sideburns and drove around a fleet of pink Cadillacs. Who? Why, Elvis was his name.

Elvis Presley came out of nowhere, recording first for Sun Records (a small firm in his hometown), with money he earned from driving a laundry truck. He captured the musical quality of Black singers at a time when there were two separate markets, one for white singers and one for Black singers. In the beginning, his performance was as much visual as vocal. He appeared on Ed Sullivan's TV show with the bottom half of the picture blacked out (showing him only above the waist) because he gyrated wildly as he sang. Some people thought the movements of his hips were obscene.

Radio couldn't take credit for creating Presley; it was TV that made him, although Presley (directed by a shrewd manager, Col. Tom Parker) charged so much for TV performances that few shows could afford him. Further, Parker limited the number of his appearances so that he would not be seen too much and lose his appeal.

55 Dinah Shore got her start on *Grand Ole Opry*.

56 Eddie Fisher made many smash hits in the fifties.

57 Roy Rogers.

58 Lulu Belle and Skyland Scottie, *National Barn Dance*, 1942. — (NBC photo)

59 Gene Autry.

Thus, radio introduced each new song to Elvis' adoring fans and each new song got plenty of air time. For a while, he seemed the only singer around. Listeners rushed out to buy Elvis' records after hearing them on radio.

Other Rock 'n' Roll Singers

The rather extraordinary thing about the singers of rock and roll is that they came and went rather fast. The singing star of years past endured (like Crosby and Sinatra), but the newer singers tended to disappear after one hit. Most of them created novelty numbers but were not experienced or talented enough to become permanently successful. Yet they were created, greeted and nurtured by the disc jockeys as carefully as the great talents before them. Rock 'n' roll groups took over radio record shows. Nothing else was heard.

Payola

Then the bombshell hit. It turned out that many disc jockeys were on the take. A new word was created: *payola.* Some record companies paid d.j.'s to play a record over and over and to plug it until it became a hit.

Even without payola, the disc jockeys were making big money. Al "Jazzbow" Collins, a respected New York d.j., claimed that any record spinner "who made less than $50,000 a year legitimately wasn't worth his salt." But others pushed their salaries higher by dishonest schemes. Many of the crooked d.j.'s caught by the investigation were fired; others resigned.

Changing Styles

Murray the K burst on the scene in the early sixties. He took over the spot vacated by Alan Freed, New York's Number One D.J. (Freed had resigned on the air during the payola scandals.) Murray the K hit it big when he attached himself to the music craze that swept America—the Beatles. (He called himself "the fifth Beatle.") When the four mop-tops from England landed on the shores of the United States in 1964, we were already primed for them. They took over. The emerging youth culture with its demonstrations, radical politics, and liberal social goals dates from the invasion of the Beatles.

The style of popular music changed so swiftly that in a matter of months almost every station devoted its record show air time to rock music. You simply had to choose among British rock, acid rock like that played by Jimi Hendrix, folk rock by Bob Dylan, bubblegum rock (The 1910 Fruitgum Co.), and soft rock (The Mommas and the Poppas). Radio was almost all music. And the lyrics dealt with ideas and themes that had not been dealt with before. However, some stations forbade the use of lyrics dealing with drugs and related subjects.

A few stations perceived a market in non-rock and gradually middle-brow stations began to play arrangements of pop rock hits done by groups which toned them down in style and beat into another form of Muzak. On all stations disc jockeys played music and seldom talked except to give commercials.

Many of the former big name singers were not heard from at all during the sixties. Indeed, many teenagers (unfamiliar with the big bands and singers of the past) seemed to believe that music had just been invented. There were many listeners who did not know that music had been written before the Beatles or the Rolling Stones and

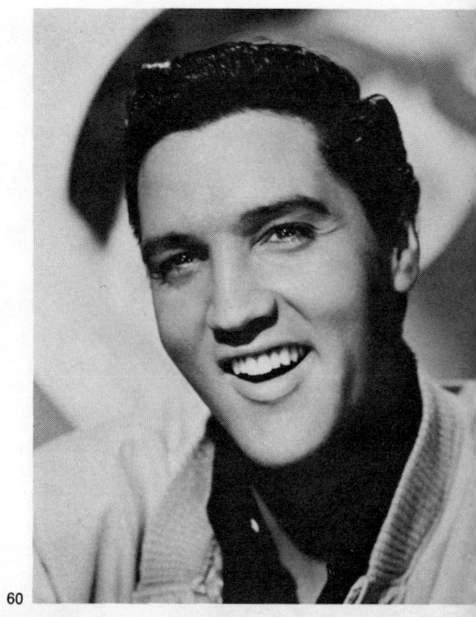

60

61

62

63

60 Elvis!

61 Singing star Tommy Edwards.

62 The popular Crew-Cuts.

63 The Beatles!

64 "Swing and sway with Sammy Kaye," introduced this popular band leader in the forties.

65 Benny Goodman had them dancing in the aisles at the Paramount theatre in New York City when he played his "licorice stick."

although they could talk about groups like Paul Revere and the Raiders or Herman's Hermits, they were ignorant of Bo Diddley and Elvis Presley, let alone Rodgers and Hart or Oscar Hammerstein.

Music Programming

In the late sixties, radio seemed to be only music. There were two ways for d.j.'s to organize their shows. One was to follow a list of the top forty tunes and play them over and over again in order. The d.j. was given a book or script to follow and he did so, reading each commercial and playing each musical selection strictly according to schedule.

The other format was called "free form" in which the d.j. selected the music he wanted to play and organized his program as he wished as long as he covered the basic commercials. This, of course, is a far cry from the days when people phoned in their requests and the d.j. dedicated the next number " 'Cry' for the gang at Rudy's Malt Shop, for Jill and Joe, for Sally and Sandy . . . "

Saturday Night Swing Club

This early disc jockey script meant the announcer couldn't ad-lib between selections.
D.J.—Mel Allen

Theme.

ALLEN.—The Saturday Night Swing Club is . . . now . . . in . . . session!

Theme.

ALLEN.—And so our theme heralds the opening of another session of the Saturday Night Swing Club . . . the ninety-fifth in a series of programs devoted by the Columbia Network to that thing called swing. . . . Brother Ted Husing's away covering the old Kentucky Derby today . . . and so yours truly, Mel Allen, will carry on for the dear old Club. . . . I'm asking you to swing it, Mel Allen. . . .

ALLEN.—Thank you, Mel. . . . Greetings, swing fans everywhere. . . . I've been on the pitching end of that introduction for just about ninety weeks now, and let me tell you it's a pleasure to be on the receiving end too . . . and the show we've got lined up for you tonight looks like a honey . . . as our guests, two of the strolling minstrels of five two boulevard, Mr. Addison and Teddy Bunn of the Onyx Club, with some two-guitar specialties that really swing . . .

ALLEN.—For the vocal division . . . Columbia's sweet young swinger of songs . . . that's right . . . it's NAN WYNN we're talking about . . . friend Paul Whiteman's sent over his new small jazz combination . . . the Whiteman Swing Wing, with Mr. Jackson Teagarden presiding on trombone . . . a little later on, Leith Stevens wants you to hear a brand-new instrumental novelty he's cut out and glued together . . . but right now . . . time for Stevens and Company to ring up the curtain with a new rhythm ditty . . . *Something Tells Me.*

1. Something Tells Me

ALLEN.—And that was *Something Tells Me* with Leith Stevens and the Swing Club doing the telling. . . . Now for something fresh off swing alley. . . . Mr. Addison and Teddy Bunn with some special string-swing on two guitars . . . and if their first tune is as wacky as the title, looks as if we're in for some typical Teddy Bunn stuff and nonsense . . . it's called . . . *Ducky Wucky.*

2 a. Ducky Wucky

Swell, Mr. Addison and Teddy Bunn . . . What have you got in the way of an encore? Oh, yes, a plug for Uncle Joe . . . *The New Onyx Club Special.*

2 b. The New Onyx Club Special

ALLEN.—Thank you, Teddy Bunn and Mr. Addison . . . really swell two-guitar stuff. . . . You know, in every band's library . . . there's an arrangement or two that is just so solid the boys would like to play it every set. . . . The Swing Club's no exception . . . we've got a couple we like to tear down over and over again . . . like the one Leith and Company are gonna get off now . . . *Rose Room.*

3. Rose Room

ALLEN.—Of all the gals around Columbia who add rhythm to a tune and really make it swing . . . the musicians' favorite is that popular favorite . . . little NAN WYNN . . . who has really big ideas about swinging a song. . . . Nan, come on over and say hello . . . and while you're here, why don't you get goin' right away on that grand blues you do so well . . . *I Can't Face the Music. . . .*

4. I Can't Face the Music

ALLEN.—Nan . . . that was really something. . . . I know a lot of our fans will want to know that you recorded that recently with friend Teddy Wilson . . . and now, Leith Stevens has something up his sleeve . . . but it isn't off the cuff . . . it's a brand new instrumental specialty in that good old Kansas City Stevens style . . . a rhythm ramble with another one of those out of the world titles. . . . Leith, let's have the world *première* of . . . *The Shadow Knows. . . .*

5. The Shadow Knows

ALLEN.—Thank you, Leith . . . that was the Stevens opus . . . *The Shadow Knows* . . . and if this particular shadow knows, you'll be hearin' it again. . . . Now . . . time for the old Father's brand-new swing combination . . . the Whiteman Swing Wing . . . with Big Gate . . . Mr. Jackson Teagarden on the trombone . . . Little Gate . . . Mr. Charles T. on the trumpet . . . our old friend Sal Franzella on the clarinet. . . .

The first tune they're gonna rip off is strictly for the gents who like their jazz Dixieland . . . and that means just about everybody who likes jazz . . . a one-a . . . two-a . . . *The Ja-hazz Me Ba-lues. . . .*

6a. Jazz Me Blues

Very swell, Jackson and Company . . . Now, Mr. T. . . . why don't you just ease over to this mike and tell your Uncle Mel what's under the Swing Wing this trip? . . . *'S Wonderful?* . . . I'll bet it is. . . .

6 b. 'S Wonderful

66

67

68

66 Deanna Durbin was a popular vocalist who appeared in the movies and on radio.

67 Major Edward Bowes, and the gong that told amateur contestants they were intolerable. —(NBC photo)

68 Glenn Miller's recordings of *Little Brown Jug* and *String of Pearls* were tremendous hits in the World War II years. He died in a U.S. Army plane crash.

1942

ALLEN.—Thank you, Jackson . . . and all the boys in the Whiteman Swing Wing . . . that's really swell jazz . . . and now, how'd you fans like another ditty by the Wynn that's known as Nan? . . . *(Applause)* Me too . . . Nan, what'll you do for us? . . .

WYNN.—How's about *John Peel* . . .

ALLEN.—Yoicks . . . view halloo . . . and let's have it. . . .

7. *John Peel*

ALLEN.—Thank you, NAN WYNN . . . swell to have you on the meetin' . . . and now, Leith Stevens and the Swing Club band have a little number in the books that's just rarin' to go trippin' on the air waves . . . so . . . *Stop . . . and Reconsider. . . .*

8. *Stop and Reconsider*

ALLEN.—This is Mel Allen bringing the ninety-fifth session of the Saturday Night Swing Club to a close. . . . We want to thank Leith Stevens for his conducting . . . and all our guests for sitting in . . . and we want you Philadelphia fans to jot down a date for next week . . . next Saturday the Swing Club will come to you from the stage of the Earle Theater in Philadelphia . . . where Leith and the gang are playing all week. . . . We hope we'll be seeing all our Philadelphia friends at the theater . . . and we hope all our fans will be sitting in at their radios.

RADIO FACTS!

Amateurs!

Singers, accordion players, tap dancers and other would-be entertainers had a chance to exhibit their talent on *Major Bowes and His Original Amateur Hour, Horace Heidt's Show,* and *Arthur Godfrey's Talent Scouts.* Although most of the talent was second rate (and one MC, Major Bowes, felt no qualms about stopping an act in the midst of the routine if it was below his standards) a few performers achieved popularity (Dick Contino, for example).

Barbasol.
Barbasol.
The brushless shaving cream supreme.
Leaves your face so smooth and clean.

GETTING THE NEWS

Chapter 8

69

70

71

69 H. V. Kaltenborn, the "Dean of Commentators."

70 Gabriel Heatter (MBS). — (Mutual Radio)

71 Quincy Howe, CBS newsman. — (CBS photo)

Flash! We interrupt this program for a bulletin . . . Radio was always connected with the news. Invented as a means of transferring information from one place to another, first by Morse Code and then by voice, radio's role as a news medium seemed obvious.

Presidential Returns

The first news broadcast was of the Harding-Cox presidential election on November 2, 1920. The returns were received by phone and broadcast immediately. So unusual was this event that people paid radio owners to give them the results over the phone. At least one motion picture theater set up a radio so its customers could hear the broadcast.

Later, a radio broadcast was made of Harding's inauguration speech.

Other News Events

Generally, in the twenties, speeches and political events were considered vital news. In 1921 Herbert Hoover, then Secretary of Commerce, broadcast a speech over KDKA in Pittsburgh, one of the pioneer stations. On June 10, 11, and 12, 1924, the Republican convention in Cleveland was broadcast. Graham MacNamee and John Andrew White reported the Coolidge nomination. Later that year the Democratic convention went on radio from Madison Square Garden. The Coolidge inauguration was broadcast on March 4, 1925, to about fifteen million people on a network of stations called a "chain." The Republican and the Democratic conventions of 1928 were also broadcast.

But events other than politics got radio coverage. In 1929 one of the most famous newsmen of all time, H. V. Kaltenborn, went on the air to discuss Byrd's flight over the South Pole.

Lucky Lindy

In May, 1927, what to many Americans was the century's most memorable feat occurred. Charles Lindbergh flew alone across the Atlantic Ocean and landed in Paris. When he returned to New York, announcer Graham MacNamee welcomed him and interviewed him. Listeners marvelled; Lucky Lindy was in their very own living rooms.

Listeners soon got used to hearing famous people speak—from Mussolini to Pope Pius to Gertrude Ederle (who swam the English Channel)—even the President. On March 12, 1933, Franklin Roosevelt broadcast his first Fireside Chat (an informal talk on governmental decisions and proposals).

Radio news was fast, too. There was an attempted assassination of President Roosevelt in Miami in 1934 and within an hour and a half there was radio news coverage, astoundingly quick in those days.

News Reports

Radio became *the* source for news. People became accustomed to hearing the news *first* on radio. Newspapers couldn't compete, even using extra editions, with the bulletins and news flashes sent out on the air minutes after the event,

FLASH! (July 26, 1934) Colonel Walter Adam tells of the assassination of Chancellor Dollfuss in Vienna!

FLASH! (1935) Bruno Hauptmann, who had kidnapped the Lindbergh child, is electrocuted. (Gabriel Heatter covers the event, adlibbing for three hours.)

(In the same year, Haile Selassie, Emperor of Ethiopia, pleads over radio for help to aid in his battle against the invading Italian Army.)

FLASH! (December 12, 1936) King Edward VIII renounces his throne to marry Wally Simpson, a divorcee from Baltimore! (The voice of the King is broadcast around the world.)

FLASH! (May 6, 1937) A German dirigible, the Hindenburg, explodes when attempting a landing at Lakehurst, New Jersey! Herbert Morrison, WLS-Chicago, is there with recording equipment (it is not yet practical to broadcast directly from remote locations). When the dirigible bursts into flames, he is so emotionally upset that he finds it difficult to speak. The resulting recording is still powerfully moving.

FLASH! (1939) War breaks out in Europe!

FLASH! (1940) President Roosevelt draws the first draft number!

FLASH! (1941) The Japanese have just bombed Pearl Harbor. (Scarcely anyone knows where Pearl Harbor is, but by keeping glued to our seats for the next few days, we find that a sneak attack has been made on one of the U.S. naval bases and that we are plunged into World War II.)

News Writing

The language of newscasts was simple, direct, informal, and brief. News writers used short words and brief sentences (to make the report easily understood) and repetition (to clarify unusual words).

Early radio newsmen found many restrictions. Censorship by sponsors and limitations by networks were common. When NBC vice president Frank Mason said in 1935, "(NBC) does not feel that it has a responsibility to its listeners to supply all the news . . . Radio is an entertainment and educational medium," he voiced the sentiments of many station owners (some of whom felt newscasts created controversy which could hurt their business). News writers (and commentators) had to walk a narrow path between offending powerful sponsors and station management and reporting uncomplimentary facts that they sometimes uncovered about business and government.

Reporting Styles

There were different styles of reporting and commenting on the news. Floyd Gibbons was one of those early newsmen. Somewhat dashing, with a white patch over his left eye (he had been shot in the head while a correspondent in World War I), he was a Chevalier of the Legion of Honor and held the Croix de Guerre. He had an adventurous life as a reporter with Pancho Villa, a Mexican general, in 1915; he was on the *Laconia* when it was torpedoed. In 1932 he broadcast news reports from the battlefields of China. On the air he was noted for his ability to speak at breakneck speed.

Not all commentators were glamorous, but they had unique styles. Boake Carter impressed listeners with his erudite British accent; traveler Lowell Thomas used short sentences in brief narratives filled with personal reminiscences; sad-voiced Gabriel Heatter ("There's good news tonight.") created a poetic news style with his repetitions and intonations; and unemotional Elmer Davis seemed precise and exact with his use of long sentences and colorless words.

Edward R. Murrow's broadcasts, which he wrote himself for *This is the News,* were clear, his sentences short and crisp—often blunt. In 1947, he began to use the services of a writer (Jesse Zousmer) when

72

73

74

72 Walter Kiernan, ABC radio commentator. —(ABC photo)

73 George Hicks, WABC war correspondent who reported the D-Day crossing. —(WABC photo)

74 William L. Shirer, radio commentator and expert on Hitler and Nazi Germany.

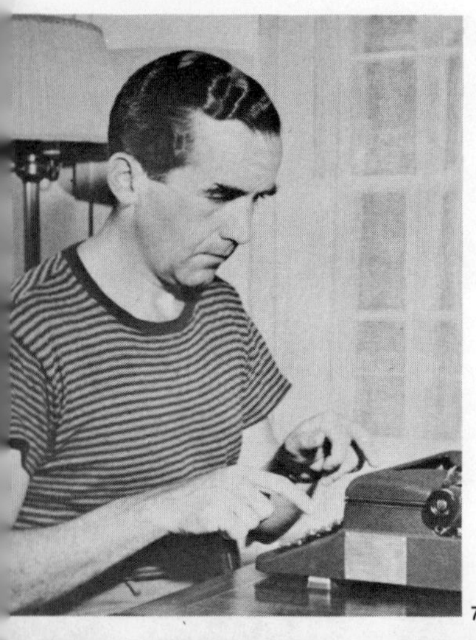

75

76

77

he did *Edward R. Murrow and the News.* Murrow's style, comprised mainly of simple and compound sentences (with no dependent clauses), was easy to follow.

William L. Shirer reported from Berlin in a conversational style. He used colloquial expressions and seemed to achieve his effect by an almost off-handed casualness.

World News Round-Up

A new idea in newscasting, using many newsmen located on the spot in different parts of the world, was created by NBC.

The first World News Round-Up was broadcast at 8:00 p.m. (EST) on March 13, 1938. The anchorman was Robert Trout in New York, with William L. Shirer in London, Edgar Ansel Mowrer in Paris, Edward R. Murrow in Vienna, Frank Gervasi in Rome, and Pierre Huss in Berlin. Their on-the-spot reporting gave a sense of immediacy and accuracy to the program.

H. V. Kaltenborn

One of the most respected newsmen continued to work alone. H. V. Kaltenborn (who spoke French, German, Spanish, and English) became a radio commentator at 43. He criticized Secretary of State Hughes for his rejection of diplomatic recognition of Soviet Russia and was fired from WEAF. At 50, he went to CBS and became famous nationally.

In 1936 he broadcast the actual sounds of guns firing in the Spanish Civil War. But his real feat was in analyzing the news (he once was reported to have interpreted a prayer). Like Lowell Thomas, he had traveled widely and was able to interject his impressions and drop names ("As Mussolini said to me . . .").

Many radio stations believe that newscasts should not contain personal opinions. Kaltenborn believed the opposite.

Winchell

Not all newsmen dealt with events of world-wide importance. Walter Winchell was a former tap dancer who evolved into what most of the country thought of as the ultimate New Yorker—brash, arrogant, and worldly-wise. He was more interested in gossip than straight news. His wide network of informants fed him all sorts of information and he put it together with clever puns, word weddings (for divorced he used "Renovating"), and slang. His Sunday evening program rattled with the sound of Morse Code telegraph keys which gave the listeners the impression that he was getting the news hot off the air as he was delivering it. Later, when he went on TV, viewers saw him tapping the key himself as the show started; the illusion was destroyed.

There were other radio gossips like Dorothy Kilgallen, Hedda Hopper, and Jimmy Fidler. All were eager to spread the latest news about who had been seen with whom in Hollywood.

Competing with TV News

Television started fifteen-minute newscasts with film coverage, utilizing the talents of such expert newsmen as John Cameron Swayze, Chet Huntley, and Walter Cronkite. Programs expanded to thirty minutes and then to one hour. Radio, down but not out, bounced back in the sixties by creating all-day-all-news programming.

75 Edward R. Murrow, celebrated newsman. — (Wide World photos)

76 The controversial gossip columnist, Walter Winchell.

77 Howard K. Smith of CBS News. — (CBS photo)

Stations using the all-day-all-news format operated just the way Top Forty disc jockeys did. They taped news and played the tapes continuously, updating them if new information arrived.

Critics have questioned whether there is enough news to fill all the time allotted to it. After all, World War II was covered in daily fifteen-minute news broadcasts (and bulletins). Those who dislike this type of news programming claim that all-day-all-news format writers and reporters—who must fill many hours of airtime—create news. That is, the reporters make small events into gigantic news. A fist fight between two kids in a cafeteria covered by TV cameras suddenly appears to be a teenage riot.

Further criticism has been made of recent attempts by news teams to joke with each other on the air. Some critics feel that kidding about the news lessens the impact and importance of newscasting.

Differences in the Modern Newscast

On most news programs the commentator has been replaced by a team of specialists who give capsule summaries of local news, world news, sports, and weather. No longer does the listener choose a particular newsman but rather tunes in a station and a news team. At present it seems that the news is entertainment, too.

Lowell Thomas

The following is an excerpt from a Lowell Thomas newscast. (March 12, 1954).

March 12, 1954—The army-McCarthy feud reached a thundering climax today. According to an army statement, Senator McCarthy tried to get an officer's commission for David Schine, an investigator for the McCarthy committee.

The Army refused, and Schine was drafted as a private. Whereupon attempts were made to get favored treatment of various kinds for Schine. There was even a plea to get Schine out of K.P. duty. Which the Army also rejected.

The army report pictures committee counsel Roy Cohn as trying to exert pressure in all this—making promises, making threats. Climaxing with menace—saying he'd "wreck the Army." He'd get Army Secretary Stevens fired!

This afternoon Senator McCarthy, with Roy Cohn sitting beside him, told a news conference that the army charges were "blackmail." He said the army people boasted they were holding Schine as a "hostage."

He says that Secretary of the Army Stevens suggested that the McCarthy subcommittee switch its investigation from the Army to the Air Force, the Navy, and the Defense Department. This is denied in the most positive terms by Army Secretary Stevens.

RADIO FACTS!

More News!

In addition to conventional newscasts, there were these programs which dramatized news and historical events: *You Make the News, You Are There. Hear It Now, The Year in Review, News of Youth, The March of Time.*

1946

1946

Look sharp!
(Bell)
Feel sharp!
(Bell)
Be sharp!
(Bell)
Use Gillette Blue Blades
With the sharpest edges ever honed!

IT'S OVER THE

FENCE

Chapter 9

Clang! The sound of the bell beginning a boxing round always brought radio listeners smartly to attention. Blow-by-blow descriptions of boxing matches thrilled sports fans. And they listened closely to play-by-play descriptions of baseball, football, hockey, and other sports.

78

79

Boxing

The first prize fight (and the first sport) to be broadcast was the bout between Jack Dempsey and Billy Miske in Benton Harbor, Michigan on September 6, 1920. Station WWJ (Detroit) carried the fight, which ended with Dempsey k.o.-ing Miske in round three. On December 22, 1920, the first broadcast direct from ringside was made of the bantam-weight match between Joe Lynch and Peter Herman.

The Jack Dempsey-Georges Carpentier fight in Jersey City, New Jersey, was broadcast on July 2, 1921. According to the famous sportscaster, Red Barber, Announcer J. Andrew White described the fight at ringside blow-by-blow over the phone to a perspiring technician, J. O. Smith, at the transmitter in nearby Hoboken. Smith wrote down what White said and repeated it over the mike. According to one report, the transmitter overheated and melted shortly after the bout ended. The broadcast was carried on WJZ (Newark) and WGY (Schenectedy). Although 90,000 people saw it in person, between 200,000 and 300,000 listeners "heard" Dempsey knock out Carpentier in the fourth round.

Boxing matches proved to be popular with radio audiences. The Dempsey-Tunney fight was carried over a 69-station chain.

Radio created its own myths and one of them was the boxer Joe Louis. In the thirties, Joe Louis fought Max Schmeling and lost, made a fight against Jack Sharkey, won, and thus came back to win the championship. In the 1938 Louis-Schmeling fight, announcers broadcast a blow-by-blow description in many languages from ringside. The Brown Bomber ruled the heavyweight ranks for many years.

There were many familiar voices doing the fight broadcasts. Both Sam Taub and Clem McCarthy did blow-by-blow fight descriptions though Clem McCarthy is better known for his horse race descriptions. Ted Husing announced many sport events, including boxing. There was also Charles "Socker" Cole, and Don Dunphy, as well as the versatile Graham McNamee.

Fight broadcasts required two announcers: one to describe the blow-by-blow action in the ring and one to provide color (a description of the fighters' backgrounds, ringside activities, and future events). Most fight broadcasts contain little or no dead air—it was always filled with the words of one announcer or the other.

Baseball

KDKA did the first broadcast of a major league ball game from Forbes Field on August 4, 1921. Soon radio devoted more time to baseball than to any other sport.

In 1921 the Giants-Yankees World Series was broadcast. The opening game (October 5) was done by Grantland Rice in New York City and relayed to KDKA. In Manhattan, the series was described by a reporter at the ball park over the telephone to Tommy Cowan, an announcer, who then repeated what he heard over the mike. WJJ (Newark) began broadcasting on October 5 with bulletins from the game.

78 Joe Louis, "The Brown Bomber," ruled the heavyweight boxing ranks for many years. Fans followed his fights — on radio.

79 Radio fans worshipped "The Yankee Clipper". (Joe DiMaggio of the New York Yankees).

80 A well-known sportscaster of the twenties, O. B. Keeler, using the mike of a portable short-wave transmitter for on-the-spot broadcasts. A caddy carried the transmitter and the man on the right carried a receiver to keep in touch with the station.

81 Graham McNamee

82 Ted Husing, Sportscaster.

83 Clem McCarthy in 1936. — (NBC photo)

In the 1925 World Series between the Washington Senators and the Pittsburgh Pirates, WHAS, Louisville, Kentucky, arranged for two local teams to act out the series play-by-play on a ballfield in the city. Over 2,000 fans each day watched the teams move around on the field re-enacting the plays that were being performed. Thus, when a line drive was hit at Griffith Stadium in Washington, at Parkway Field in Louisville, a player dutifully swung the bat, hitting an imitation line drive, and sprinted off towards first base.

At one time, announcers at some stations attempted to recreate a ball game by reading the telegraphed accounts of the game. Some even added crowd noises and other sound effects to give the impression they were broadcasting from the game.

Gradually, baseball announcers became famous. In the days when the announcer sat in the stands with the rest of the fans, some sportcasters even became legends.

Graham McNamee, one of the best known early sportscasters, with Bill Manning (the original "Old Redhead") broadcast the 1929 World Series. Red Barber started with Cincinnati as an announcer in the thirties, then went to the Dodgers and then to the Yankees. Bob Elson started with the Cubs in the thirties and then went with the White Sox. Mel Allen became the voice of the Yankees (He always said "How about that?"). Russ Hodges worked for the Giants.

Baseball was slow moving; there was a lot of empty air time between pitches and usually this was filled detailing statistics, giving information, and describing the players and the crowd.

There were other announcers, many who were former baseball professionals: Mel Ott and Leo Durocher, who broadcasted for the Giants, Curt Gowdy, Frank Frisch, Arch McDonald ("The Old Pine Tree")—a well liked phrasemaker (He named Joe DiMaggio "The Yankee Clipper"). The most unforgettable was Dizzy Dean (who once played for the Cardinals). Dean mangled the English language, using words like "slud" for "slide." Ex-baseball players made good sportscasters. You want evidence? Pee-Wee Reese and Joe Garagiola are good examples.

Football

Fall Saturday afternoons were devoted to broadcasts of college football games. (It wasn't until television became available in the fifties that pro-football captured the public's interest.)

The first play-by-play broadcast of a football game was the November 25, 1920 Texas U.-Texas A&M contest on 5XB in Morse Code. The first coast-to-coast voice sports broadcast was the Princeton-Chicago game on WEAF (New York) on October 28, 1922. The first coast-to-coast broadcast of a Rose Bowl Game was the one between Stanford and Alabama on November 29, 1927. Graham McNamee did the broadcast.

Football Sportscasters

Ted Husing was the first sportscaster to introduce the spotting board, containing a list of the names of the players, so that an assistant could silently point out the name of the player making a tackle or catching a pass. Husing's board was fancy—it lit up. Not wanting to reveal the tricks of the trade, he kept it concealed from his fans and others.

Bill Stern (NBC) was often criticized because (it was claimed) he was inaccurate. In football, his critics said when Stern erred in identifying the ball carrier, he covered himself by saying the ball was then passed to the correct player.

Other sports announcers were Bill Munday, best known for his colorful descriptions of football, and Bill Slater (MBS), a sports announcer who also did *Twenty Questions.*

Radio Football

Football, with its time limits, created a different effect on its listeners than open-ended baseball, which could run on for hours. Somehow the sound of thumping drums and the cheering crowd conveyed a poetic picture in the imagination.

Listening to sports programs was somewhat different from listening to other radio shows. For example, it was possible to listen to fantasy or science fiction, imagine a world that did not exist, and create it in your mind. In listening to a radio broadcast of a sports event, the listener had to have some idea of the rules of the game, what the playing field looked like, and the rituals of the game. Sports programs were good entertainment because no one could predict or figure out what was likely to happen because chance ruled it all.

There were heroes to follow (Babe Ruth, Joe Louis, for example). Each of the shows had a plot—the favorite team was the protagonist and the opposing team was the antagonist; there was a winner and a loser.

Although some sports were inexpensive to broadcast, there were problems. For example, if weather interfered (and a game was rained out) substitute programming had to be immediately available. Stations regularly used recordings for this purpose although in the early days there was a singer and pianist paid to stand by. Too, the announcer had to be able to glibly ad-lib, that is, be able to talk at length on almost anything, but not to get too excited, swear, or say anything off color. One famous announcer uttered profanity on the air in later years and it brought his career to an instant end.

The announcer had to remember that his audience could not see the game and he had to describe it clearly. A master in understanding his audience and their needs, Red Barber used an egg timer to remind him to mention the all-important score every three minutes.

As the "Old Redhead" points out, it is vital for the sports announcer to be impartial, well prepared, aware of the stakes, able to concentrate solely on the game, ask why things happened, and be unflappable.

Back on August 4, 1921, KDKA broadcast the Davis Cup Matches. Later, horse races went on the air (with Clem McCarthy or Jack Drees). By the forties, every sport seemed to be broadcast on radio: horseracing, baseball, football, golf, hockey and even chess.

Sports broadcast on radio created myths. The sports figures described on radio existed in the listeners' imaginations—thus, they were bigger than life.

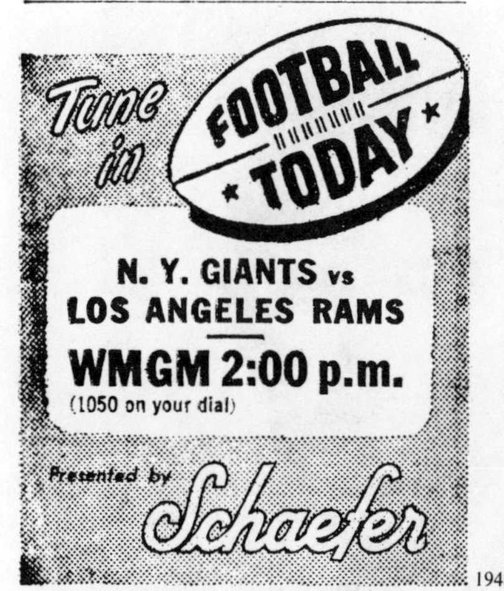

1948

EXCLUSIVE FOR

SPORTS FANS

a preview of the 12 major league ball teams now training on the East Coast are all covered on

BETHLEHEM STEEL'S

RADIO PROGRAM

6:30 P.M. • WJZ

MONDAY THRU FRIDAY

HARRY WISMER is the only national radio sports commentator to cover the camps in person, the only commentator in any branch of sports to see all the 12 eastern camps.

Listen to his comments daily, Monday thru Friday, and don't miss his sum-up on March 29th.

1946

1943

84 The great sports announcer, Bill Stern

Army-Notre Dame Football Game
Second Quarter, November 7, 1942

This transcript is from the play-by-play description by Bill Stern broadcast over the Blue Network.

. . .The ball goes over to the Army and that is the third time today that Notre Dame has failed to score . . . when they were within scoring distance. Troxel comes back in at fullback and Roberts comes back in at quarterback for the Army . . . At the end of the first quarter, Navy leads Pennsylvania 7 to 0; the Pennsylvania team that defeated the Army last week 1 to 0 and the same Navy team that Notre Dame defeated last Saturday 9 to 0 are playing today in Franklin Field, Philadelphia and Navy leads 7 to 0 . . . All right . . . the Army team has Roberts, Maiser, Troxel and Hill in the backfield. It's a single wingback over the left with Hill leading. Hill takes a pass at center, breaks into the clear, smashes down to the end of the 33 yard line before he is stopped by Louie Rimkiss . . . A gain in that play of 8 yards . . . 8 yards, second down; one; that was Hill No. 17, the right half-back of the Army, smashing down to the 34 yard line of the Army before he was finally stopped . . . The ball 15 yards in from the east side of the field . . . He was tackled by Clayton Miller, the right half-back of Notre Dame . . . Out of the huddle comes the Army team, . . . single wingback off to the right, Maiser in the full back position. The ball goes to Henry Maiser who gives it to Callaher; . . . Callaher is running wide; he is up there and knocked out of bounds That was the end of round play . . . the end of round play with Callaher carrying the ball loses yardage for the Army. They lose four yards on the play, back to the 30 yard line . . . The tackler was Joseph Lamonte . . . So it's third down, about 3-1/2, Army's ball, 15 yards in from the west side of the field, the score nothing to nothing in the second period . . . This is the Blue Network presentation of the game of the day from Yankee Stadium New York City . . .

(Drum beating to keep time with each letter.)
L-A-V-A.
L-A-V-A.

RADIO FACTS!

Novel Sports Shows!

In addition to sportscasters describing games play-by-play, there were sports quizzes (*We Want a Touchdown*, a football quiz with Red Barber as MC), recreated old-time sports events (The Liberty Broadcasting System was the source), and dramatized life stories of famous sports stars (*The Joe DiMaggio Show*).

WHATEVER HAPPENED TO RADIO?

Chapter 10

NEW FM STATION BEGINS OPERATION

NBC Outlet Installed On Empire State Tower.

The "green light" given frequency modulation by the FCC has spurred on the industry to greater efforts in meeting the expected demand for transmitters and receivers. On Thursday night, the National Broadcasting Company threw the switch on its FM station, W2XWG, in the Empire State tower thereby increasing to five the number of transmitters on regular schedules in the metropolitan area.

At the same time, rumors from set manufacturers tell of many new FM models which will be announced during the next six weeks.

The NBC station is a 1,000-watt unit built under the supervision of Raymond F. Guy of the engineering staff and designed to use the standard 150 kilocycle deviation. The antenna in use is a temporary one but eventually the FM signals will be radiated over the antenna erected on the tower for television, the two services operating simultaneously.

For the present, W2XWG will transmit regular NBC programs from 4 to 11 P. M. on Tuesday through Saturday of each week, these times corresponding with the television schedule.

Tests made on Thursday night showed that the FM programs from the Empire State tower could be received with adequate signal strength at a distance of 25 miles or more and without interference from other FM stations. W2XWG operates on 42.6 megacycles.

Other frequency modulation stations in New York city are W2XMN (42.8) the pioneer station erected and operated by the inventor of FM, Major Edwin H. Armstrong; W2XQR (43.2) operated by John V. L. Hogan; W2XOR (43.4), sister station of WOR; and W2XWF (42.18) operated by William G. Finch. Occasionally, W2XMN also transmits simultaneously on 43 megacycles.

1940

At the end of World War II, TV sets first began appearing in stores and salesrooms. At that time they had tiny screens ranging in size from three inches to seven inches; for a long time twelve-inch models were advertised as giant screen sets. The family who owned a TV sat in their darkened living room crowded with visiting neighbors watching ancient British films and wrestling. But that was enough. Radio seemed doomed. People were even willing to watch test patterns.

Radio sales dropped, as did attendance at the movies. Some radio programs made the switch to TV without much change, as, for example, *The Goldbergs* (in 1949) but many could not make the switch at all. *Lights Out* and other adventure shows were less frightening on television. The stalwart *Amos 'n' Andy* show moved to TV and substituted Black actors for Gosden and Correll (who sometimes did the radio show in blackface). It didn't work. Protests by civil rights groups drove the program from the air.

Radio programming A.T. (After Television), consisted entirely of records. Disc jockeys took over. There were fewer and fewer dramatic shows. Comedy dwindled and disappeared. Sponsors wanted the most for their money and that appeared to be TV. Actors sought out jobs in TV, hesitantly at first, and then plunged in. It looked like radio had been done in by TV just as vaudeville had been done in by radio.

But there was a lot of money invested in radio equipment and talent and not everybody wanted to see it wasted. Networks created new ideas and methods. *Monitor* was such a program. First conceived as a means of offering weekend entertainment, it was a curious hodge-podge of music, reports, humor, and news. The programming was very flexible and this was part of its charm; the listener never knew what to expect and he kept tuned in so that he didn't miss anything. Further, there were teasers (mention of things to come) to attract and hold the listener's attention.

Gab . . . Gab . . . Gab

TV usually ended at midnight or thereabouts while radio blasted all night long. Although much of the programming consisted of recordings, there were programs like those of Barry Gray and Jean Shepherd who talked all night, more or less, on various subjects. They filled the needs of insomniacs and workers who listened to the radio at night, but also they created a format which later was used on daytime radio and TV as well—the talk show.

Thus, the style of radio began to change. From the major source of edifying entertainment and instant information, radio seemed to have been given way to the one-eyed boob tube. There was still a place for it—though listeners no longer talked about what had been on the radio the night before, radio still served the function of imparting news, weather and musical entertainment to drivers and others who couldn't watch TV.

The really radical change was in the creation of the open-mike talk shows on which people could phone in and presto—they were radio personalities (if only for a moment). A phone patch (an electronic hook-up between phone and mike) permitted them to speak on the phone but an eight-second delay (a device which kept their voices from going out on the air the instant it was spoken) was needed to avoid the obscene caller or something going over the air that shouldn't.

One of the earlier phone-in-talk shows was Brad Crandall's on WOR (New York). He took calls from eight to ten p.m. Although he did not announce a topic, people tended to talk about the same things and in this way the program hung together. Crandall was usually a sympathetic listener but not above telling disagreeable people off.

The format was adapted to other shows. People could phone in about things that they wanted and electronic want-ads were created. Sports phone-in programs were created.

The phone-in idea was a strange reversal of the days when the radio station phoned contestants to see if they could guess the answer to a contest question.

Out of this came some shows which seemed to be X-rated or for adults only. This type of show gained popularity on the West Coast and then spread to the East. It consisted generally of people phoning in and discussing intimate sexual and personal problems and activities which many people believe should not be discussed on a public information carrier such as radio. Although they could, of course, turn the material off, the question appears to be a larger one of censorship.

News

Another change has been the approach to news. Stations had used a variety of styles; some resorting to a hoked-up news image with clacking teletype machines, and the staccato of a telegraph key to make every news event sound like a flash (just as Winchell had done). Others reverted to a formula of leisurely discussing the news as did the earliest news commentators.

But most interesting of all was the development of the all-day-long-all-news shows. In New York City, WINS, which had been a Top Forty rock station with Alan Freed, converted to the all-news format.

Music Programs

Music formats have changed to either free form or carefully programmed approaches. But the stations still retain a particular image rather than develop a series of programs for different images and different audiences.

Since music numbers are no longer confined to the straight-jacket of three minutes, but vary in length, the programming varies too.

A major change is in the number of commercials. Although there was a time when there were two commercials in fifteen minutes, there are now three or more a minute, and more are packed in over that fifteen-minute period.

Return of Drama Shows

As a result of the public interest in nostalgia in the late 60s, some stations brought back vintage radio dramas by playing transcriptions (giant 16 rpm discs) or tape recordings of *Fibber McGee and Molly, The Lone Ranger, Gangbusters,* and *The Green Hornet.* A small industry grew up selling tapes of old shows. CBS began a new radio drama series running 30 minutes five nights a week. Rod Serling wrote a series for another network. Fans began to turn off the boob tube for a while each night and listen—eyes closed—just the way people in radioland did years ago.

1948

1949

Modern Times

Obviously, modern programming has come a long way from the old days when radio nuts stayed up half the night to hear the call letters from a transmitter a couple of hundred miles away. Already we have held transistor radios in our hands and listened to men talking on the moon.

Ah, Radio, what next?

Jean Shepherd

This free-association monologue is representative of what one highly creative modern radio artist does on the air. Shepherd, who has published numerous works in addition to making recordings, appears nightly on New York's WOR. His work is poetic yet earthy.

CHANNEL CAT IN THE MIDDLE DISTANCE

But here it is on a quiet February night time to listen and time to sit, time to wait for the next channel cat to make the bend, time to wait for the next starfish to reproduce its kind, almost impossible to kill a starfish, you know, almost impossible even to understand a starfish. I once knew a starfish living just outside Hamilton, Ontario.

And if we cared to note we would see a tiny figure tattered and torn who seems to be having a few of the difficulties that he called upon himself and a few extra, too but then the machines were all well-oiled, you see a few of them needed extra bearings an occasional headgasket was there to be replaced, and a few of the valves needed grinding, too, but these things all in good time all indeed in good time it takes a little and it takes two.

Jean Shepherd: records, transcriptions and it really does take that you know it really takes that and a few things extra, too.

The whole thing is absolutely ridiculous as far as I'm concerned. This is the first time I've been ill really the first time I've spent any time at all concerning myself with medicine but nevertheless the business of being ill is an extremely interesting one and for those who haven't tried it for some time, there are certain things you should pick up before you make a serious attempt to create a small one for yourself getting back into business for yourself after you have been ill for awhile is much more interesting than being ill itself to begin with, a large number of people didn't even know you were any place at all disappointing an extremely disappointing thing I remember seeing a cartoon Charley Brown on telephone calling girlfriend at party saying he's ill and won't be able to come to the party girlfriend answering: that's O.K., didn't even know you weren't here Charley Brown hanging up telephone, looking quietly into the middle distance there are many things to be seen in the middle distance didn't even know you were gone . Monday night, yes, it's a Monday, sort of a Monday night off into the middle distance, strive and continue to strive, the three of them, each one of whom has pulled the shortest straw of the group.

And again its the golden touch the inescapable, the always present, the all-encompassing golden touch a few years ago, going out into left field, the old man made the mistake of always those in the dug-out allowing them the brief luxury the realization the knowledge that he wasn't going to make it you see, you've got to always give them the impression—you've always got to keep it up keep the footprints high, well up on the front of the cushion keep them moving.

Again the usage of the number 3 3A Sable Brush future use, future reference car size itself is most important the stories and the scores upon scores of them that were told in the time that passed were not necessarily stories of truth nor were they always stories just a few words an occasional period and once in awhile a CAPITAL letter.

And while all the thoughts were being compiled and all the words were being put down on 3 x 5 file cards a few typewriter ribbons were being changed and occasionally someone took the time to oil the fielder's mitt there were things to be written, things to be said and things to be done and there was a quiet Monday night a few people were listening a few people were looking a few stars were being watched a few moons were being examined a few sand dunes were being understood as, in fact, had ever been understood in their brief period of sitting before all the elders. It came to pass that the writing was good and we've been reading and we've been looking and we've been cleaning windows, washing glasses and smoking cigarettes and from time to time changing typewriter ribbons. We have been clipping fingernails and we've been listening to recordings. We have been eating hamburgers and we've been eating auriomycin. We've been doing a thousand things and there have been one hundred moments that have passed and gone some of them were good and most of them were forgotten and that's the most important of all, I suppose most of them were forgotten just beginning things that passed. There seems to be too much smoke and it also seems the rugs are a little threadbare the walls are not painted the right color and that's another one of the small troubles.

This is their way of going their way of disappearing and this is the thing students ten thousands years from now are really going to work upon going to understand only vaguely and are going to make definite attempts to re-create but these attempts are just exactly that just attempts. I think this is going to be one of the wonderful things we have to offer above all things. This is probably a much recorded sentence and when I speak of a sentence I mean a sentence in man's eternal struggle to create a chapter of himself rather than a paragraph, a line or perhaps a phrase but this particular sentence this particular sentence which could be called a century this time this period has been recorded ten million ways they've even recorded the disadvantages of keeping bananas in a refrigerator they've recorded the advantages of liver stimulants of one kind or another and the voices of presidents and cabinet ministers they've also recorded the voices of little people speaking to one another on the street and there shall be a few of these conversations discovered, too and it shall be the moment of discovery that will prove the small turning point in many a small scholar's career this moment of discovery that was the way these people spoke this is the way they talked and this is the way they sounded and really this is the way they were and a small partial moon went scooting out over the curving river it picked up speed and it picked up momentum and it picked up a few forgotten thoughts as it moved a small partial moon just a part an example of life an example of shadow an example of creative

RADIO FACTS!

FM Improves Music!

FM (Frequency Modulation) produces very clear reception (as opposed to AM — Amplitude Modulation — which is susceptible to static) and better quality high fidelity sound.

Invented by Edwin Howard Armstrong in 1933, FM was a boon to listeners to serious music. The first stations broadcast programs uninterrupted by commercials.

When stereo-broadcasting became technically possible, listeners became more sophisticated. The first stereo broadcasts used AM for one channel and FM for the other channel; you needed two radio sets to listen. With transmission of the two channels together on one signal on FM, the fantastic sound thus created boomed the sales of FM receivers.

thought but it was just partial, you see it wasn't complete it never would be.

We have many nutcrackers to work with but no small picks which leads us exactly nowhere because that's exactly where we intended to be there's no waiting and no wishing step right up and 7 times over, 3 times down there are 17 chairs, absolutely no waiting everybody will be taken care of in time All Ice Cream Manufactured on the Premises.

It was a good, easy Monday night the crowds were even more interesting than usual but the old man was coasting more or less—coasting in between and in betwixt what used to be called a liquid or even a solid in suspension, floating between the surface and the bottom sort of in between just floating in a colloidal suspension I had difficulty with that one time and I was ready to surmount the difficulty and here we are floating again and the surface glimmers and sheens there's much to be said for surface tension, but that, too, will have to wait till after one o'clock when Grandma has put her knitting away and gone to bed. Yes, there will be a full field inspection at 0300.

1949

1948

A Bibliography

Picture Histories

Lackman, Ron. *Remember Radio* (New York: Putnam's Sons, 1970).
A brief but interesting look at vintage radio.

Mitchell, Curtis. *Cavalcade of Broadcasting* (Chicago: Follet Publishing Company, 1970).
A useful discussion of the technology, the struggles, and the future of radio as well as the stars and personalities involved.

Settel, Irving. *A Pictorial History of Radio* (New York: Grosset and Dunlap, 1967).
A well written and heavily illustrated story of old-time radio with clusters of excerpts from the scripts of ancient shows.

Guide to Programs

Buxton, Frank and Bill Owen. *The Big Broadcast, 1920-1950* (New York: Viking, 1972).
This magnificent book lists the casts of a fantastic number of old-time radio shows and gives interesting and informative information on each. The single most important reference work for anyone interested in old radio!

Guide to Scripts

Poteet, G. Howard. *Published Radio, TV, and Film Scripts: A Bibliography* (Troy, New York: Whitston Publishing Co., Inc., 1975).
A listing of radio scripts available in published form, arranged by the name of the show.

Guide to Technical Information

Abbott, Waldo and Richard L. Rider *Handbook of Broadcasting* (New York: McGraw Hill, 1969).
An excellent overview of fundamentals of techniques, production strategies, and business activities.

Bliss, Edward, Jr. and John M. Patterson. *Writing News for Broadcast* (New York: Columbia University Press, 1971).
A useful book for gaining insights into writing for radio with examples from both TV and radio.

Coleman, Ken *So You Want to Be a Sportscaster* (New York: Hawthorn Books, 1973).
Some helpful advice for the would-be sports announcer.

Hilliard, Robert L., ed. *Radio Broadcasting* (New York: Hastings House, 1967).
This valuable volume discusses operations and production, and is full of useful information on writing and performing.

An Authoritative History

Barnouw, Erik *A History of Broadcasting in the United States; Vol. I, A Tower of Babel, Vol. II, The Golden Web, Vol. III, The Image Empire* (New York: Oxford University Press, 1966-1970).
The most accurate and complete history of radio available. A must for any vintage-radio fan.

Index

ZENITH